643.7
M - 7

McConnell, Charles.
   Building an
addition to your home.

88-26

Please return on or before the date due.
You are responsible for all materials bor-
rowed on your card. All slips must remain in pocket. A charge will be made
for any missing slips.

HENNEPIN COUNTY LIBRARY
MINNESOTA 55435

GEMCO

During his forty years' experience in construction, Charles N. McConnell has designed, installed and supervised plumbing and heating installations in residential, commercial, industrial and public buildings. He built his own house using the construction methods described in this book and has written many other books on related subjects.

BUILDING AN ADDITION TO YOUR HOME

Charles N. McConnell

Line Drawings (not otherwise credited) are by
Jerry Forney

Prentice-Hall, Inc., Englewood Cliffs, New Jersey 07632

Library of Congress Cataloging in Publication Data

McConnell, Charles N.
    Building an addition to your home.

    Includes index.
    1. Dwellings—Remodeling.   2. House
construction.   I. Title.
TH4816.M28   643′.7   82-508
ISBN 0-13-086009-3   AACR2
ISBN 0-13-085993-1 (pbk.)

This book is available at a special discount when ordered
in large quantities. Contact Prentice-Hall, Inc., General
Publishing Division. Special Sales, Englewood Cliffs, N.J. 07632.

10 9 8 7 6 5 4 3 2 1

Printed in the United States of America

ISBN 0-13-085993-1 {PBK.}

ISBN 0-13-086009-3

Prentice-Hall International, Inc., *London*
Prentice-Hall of Australia Pty. Limited, *Sydney*
Prentice-Hall of Canada Inc., *Toronto*
Prentice-Hall of India Private Limited, *New Delhi*
Prentice-Hall of Japan, Inc., *Tokyo*
Prentice-Hall of Southeast Asia Pte. Ltd., *Singapore*
Whitehall Books Limited, *Wellington, New Zealand*
Editora Prentice-Hall do, *Brasil LTDA., Rio de Janeiro*

Dedicated to my wife, Joyce, and my three sons: Harry, Jack and Charlie, whose support has never faltered.

Contents

Section Three: HOME IMPROVEMENTS

Section Four: GENERAL INFORMATION

Acknowledgments

I wish to thank the following companies for their assistance in furnishing information, photographs and drawings of their products:

Armstrong Cork Company
P.O. Box 3001
Lancaster, PA 17604

Dow Chemical USA
2020 Dow Center
Midland, MI 48640

Florida Tile, Sikes Corporation
P.O. Box 447
Lakeland, FL 33802

Georgia-Pacific Corporation
900 S.W. Fifth Avenue
Portland, OR 97204

Intermatic Incorporated
Intermatic Plaza
Spring Grove, IL 60081

The Majestic Corporation
245 Erie Street
Huntington, IN 46750

NuTone Division
Madison and Red Bank Roads
Cincinnati, OH 45227

Owens-Corning Fiberglas Corporation
Fiberglas Tower
Toledo, OH 43649

QuakerMaid, a Tappan Division
State Route 61
Leesport, PA 19533

A. O. Smith Corporation
Consumer Products Division
P.O. Box 28
Kankakee, IL 60901

Southeastern Aluminum Products
P.O. Box 4816
Jacksonville, FL 32201

GTE Sylvania Incorporated
Standard Distribution Products Division
750 Boling St.
P.O. Box 2431
Jackson, MS 39205

I also wish to thank Mr. Dennis Fawcett, senior editor, for his helpful suggestions and encouragement in the preparation of the manuscript.

Charles N. McConnell

Preface

As you read this book it will be as if you were working side by side with skilled tradespeople in building an addition to a home. The intent is to show and explain to you the principles and basic methods used in home construction. When you know the basic methods you can adapt this knowledge to your particular needs. You'll learn how to lay out the foundations, how to form and pour footings, frame walls, install electrical wiring, work with PVC pipe, solder copper tubing and much more. You'll learn how to apply and repair drywall, install ceramic tile and wall paneling and make common household repairs.

Energy-saving tips and information on energy-saving equipment and devices are included because saving energy also means saving money. Federal tax credits are being given for the installation of solar water-heating systems; this is explained in the section on solar-water heating equipment.

The suggestions on maintenance can save many costly repairs and items suggested for home improvements should add much more than their cost to the value of the home.

Building an Addition to Your Home will be a valuable reference book for your home library.

Every effort has been made in the preparation of this book to ensure that it contains correct information. However, the publisher and author assume no responsibility for errors and/or omissions and no liability is assumed for damages resulting from the use of information contained in this book.

1
NEW
CONSTRUCTION

Chapter One

Planning the Work

Before starting any building project some degree of planning is essential. Partitioning a basement or finishing an attic, as an example, will not require a blueprint to work from, but if sketches are made and referred to as the work is being done the doors, closets, etc. will be in the right place when the job is finished. If you are considering building an addition to your home you would be wise to have an architect draw the plans. The architect will be familiar with local building codes and regulations and will design the building to comply with them. The plans need not be extremely detailed or fancy: if you explain to the architect what you want to build, he or she can furnish plans showing the essential details—footings, foundations, room sizes, window and door openings, front and side views of the finished building—for a fraction of the finished cost of the addition. These plans should save you more than their cost in time that might have been spent in tearing work out and doing it over. The architect's plans won't show you how to lay out a building foundation or build a partition corner post—that type of information is in this book—but a good set of plans will show the necessary details for construction.

Chapter Two

Footings and Foundations

The footing is the base upon which the building is built. Building codes will regulate the size and depth of the footing depending on the area, the soil conditions and the height and type of the finished structure. In general, footings should be a minimum of 8 inches deep and 16 inches wide and in cold areas of the country the top of the footing should be below the frost line, or the average depth of frozen ground in winter.

Most building codes require reinforcing rods in footings: the rods are supported on "chairs" (wire supports) until the concrete is poured, and where rods are joined they should be tied tightly together with tie wire, as shown in Fig. 3. Your architect or the building inspector in your area can tell you the size and number of rods required for your footing as well as the width and depth of the footing required. Reinforcing rod and tie wire are available at building-supply stores.

When staking out an addition or a new building, care must be taken to ensure that the footings and foundation will be square. The first step in the construction of our 20-foot 0-inch × (by) 24-foot 0-inch addition will be to lay out the center lines of the footing. The addition will be built on a crawl space as shown on the footing and foundation plan. To lay out the footings and foundation we will work from the plans shown in Fig. 1. Nails should be driven into the walls at points A and C when these points are established. First, we measure out 23 feet 8 inches (24 feet 0 inches minus 4 inches for ½ concrete block) from the existing building wall at points A and C to find the center line of the footing and the concrete block wall and drive 2-inch × 2-inch stakes at B and D. At this time, hit the stakes only two or three times, just enough to keep them in the ground. Next, measure between the stakes B and D and set them to the correct measurement, 19 feet 4 inches. Drive an eight-penny nail in the center of each stake, then measure the distance diagonally as shown in Fig. 1 between the nails at points A and D and B and C. If the two measurements are not the same, the staked-out

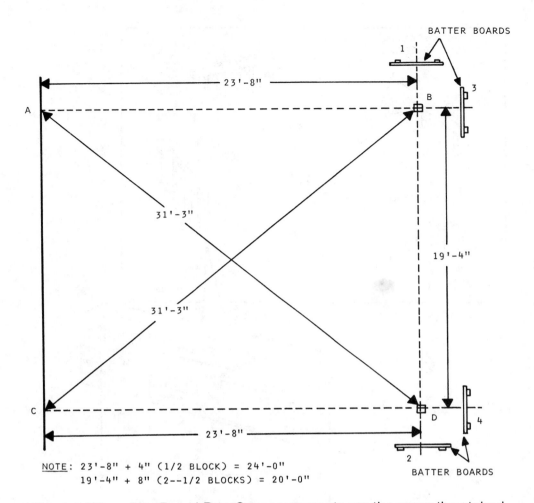

Fig. 1   When A-to-D and B-to-C measurements are the same the staked-
out area is square.

area is not square. To square the foundation the stakes at B and D
must be shifted, keeping the 19-foot 4-inch distance between them
until the measurements are identical, as shown in Fig. 1. When the
foundation has been squared off a nylon chalk line stretched from
A to B to D to C will establish the center line of the footing excavation. When the center lines have been established, drive stakes B
and D firmly into the ground. Batter boards, shown in Fig. 2,
should be set about 6 feet back from corners B and D.

Nylon chalk line can be stretched between the batter
boards and to points A and C to again establish the center line of
the wall when the footing is formed and the concrete blocks are
laid.

The batter-board stakes should be set solidly and protected

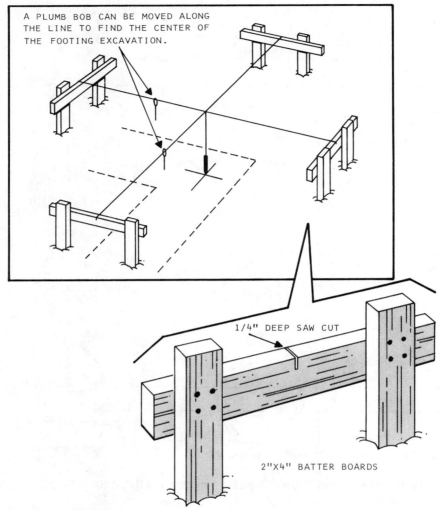

Fig. 2   Batter boards and how they are used.

during the building construction. After stakes B and D are located correctly, a string stretched across the nails at B and D to the batter boards 1 and 2 will show where the saw cuts should be made. A string stretched from the nail at A across the nail at B will show where to make the cut at batter board 3, and from the nail at C across the nail at D, where to make the saw cut at batter board 4.

When the strings are stretched across the batter boards as shown in Fig. 2, a plumb bob hanging from the intersection of the strings will show the centers of the wall lines.

EXCAVATING AND FORM SETTING
Assuming the frost line to be 30 inches deep, a ditch must be dug wide enough to work in to set forms, pour concrete and lay con-

crete blocks. This will require an excavation about 3 feet wide. An excavating contractor with a small backhoe on a tractor can dig the ditch, level the bottom ready for setting form boards and later return and backfill around the building and spread out or haul away any extra dirt.

The form boards should be 2 inches × 8 inches as shown in Fig. 3, but with today's high lumber prices you may elect to use 1-inch × 8-inch boards instead. Two-inch × 2-inch or 2-inch × 4-inch stakes will be needed to hold the forms in place while the concrete is being poured. If 2-inch × 8-inch forms are used, the stakes can be set at 5- or 6-foot intervals; if 1-inch × 8-inch forms are used, stakes will need to be set every 3 feet. The stakes are set on the outside of the forms, and ends of form boards should be nailed to stakes.

Fig. 3   Details of footings.

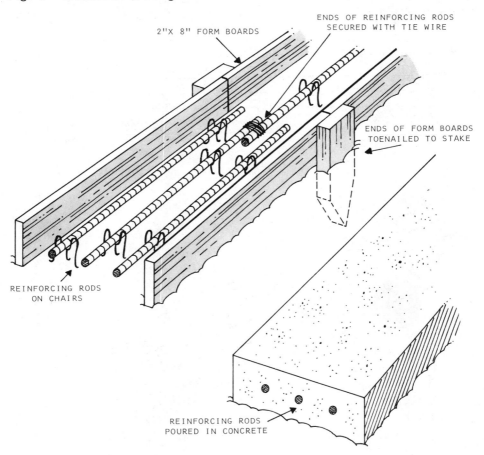

2"X 8" FORM BOARDS

ENDS OF REINFORCING RODS
SECURED WITH TIE WIRE

ENDS OF FORM BOARDS
TOENAILED TO STAKE

REINFORCING RODS
ON CHAIRS

REINFORCING RODS
POURED IN CONCRETE

For the footing shown in Figure 5, approximately 140 lineal feet of form boards will be needed. Cut two or three 2-inch × 4-inch spacers 16 inches long to hold between the forms while setting them and driving the side stakes.

After the ditch is dug, stretch the chalk line between the batter boards and drop a plumb bob from the chalk line to find the center of the forms. Stakes 2 × 4 inches should be used where the ends of the forms meet, as shown in Fig. 3. Use a 48-inch level when setting the forms, level each form board as you nail it in place, then go back and, starting at one end, check each form board again before pouring concrete. If the footing is level it will be much easier to lay the concrete blocks later. The reinforcing rods should be set in place on chairs or bricks after the form work is completed.

STEP FOOTINGS

If the ground slopes sharply at the building site, a step footing, shown in Fig. 4, will save on excavation work and also on concrete

Fig. 4   How to dig and form a step footing.

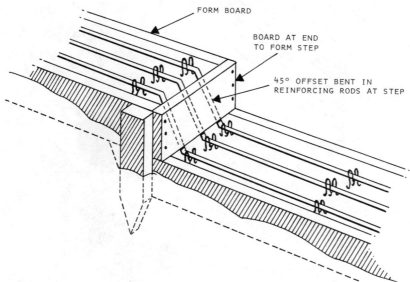

blocks and labor. The reinforcing rods should be bent to an approximately 45-degree offset and continue unbroken through the step. When pouring concrete for a step footing, you should pour the lowest sections first and allow them to set for a few minutes before pouring the top sections.

Figuring the Amount of Concrete Needed

Concrete is ordered by the cubic yard. The formula for estimating concrete is: length (feet) × width (feet) × depth (feet) ÷ 27 = volume (cubic yards) (There are 27 cubic feet in 1 cubic yard.) For this example we need to know how many yards to order for a footing 64 feet 4 inches long × 1 foot 4 inches wide × 8 inches deep.

64.33 feet (length)	85.56	2.09 cubic yards
×1.33 feet (width)	× .66 (depth of 8 inches)	27)56.47
85.56 square feet	56.47 cubic feet	

Fig. 5   Footing and foundation plan.

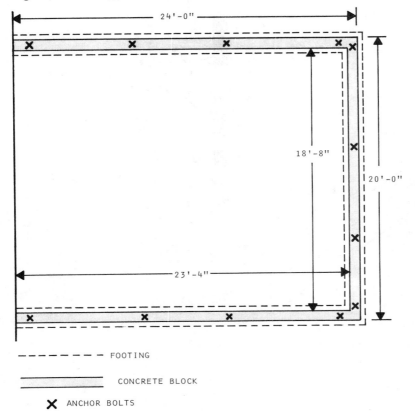

24'-0"

18'-8"

20'-0"

23'-4"

---------- FOOTING

========== CONCRETE BLOCK

✗ ANCHOR BOLTS

As shown, slightly over 2 yards of concrete would be needed for a footing 64 feet 4 inches long × 1 foot 4 inches wide × 8 inches deep. Since there is always some waste, 2.5 (2½) yards should be ordered. A ready-mix supplier can advise you on the proper strength to order. Three-thousand-pound (test at 28 days) concrete would be adequate for a one-story building in most areas. Before your concrete is delivered lay runway boards (if needed) for your wheelbarrow to run on. When concrete is delivered it is usually necessary for the truck driver to add water to the mixture in the truck to allow the mixture to slide down the chute. Add *only* enough water to permit the concrete to flow freely: concrete loses strength when too much water is added. Dump the concrete directly into the forms and "puddle" it by working a stick or rod up and down as shown in Fig. 6. This will ensure that the concrete works well down into the forms, leaving no voids or "honeycombs." Use a board as shown in Fig. 6 to drag off, or screed, the concrete level with the top of the forms. After the concrete has set up at least 24

Fig. 6   Leveling the concrete in the form.

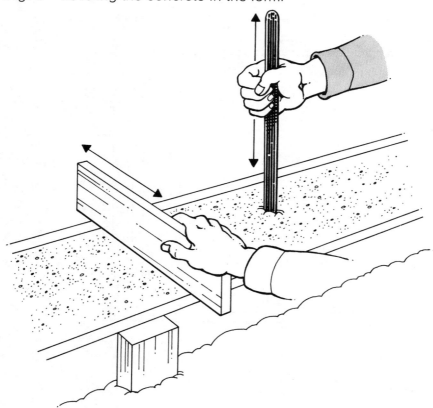

hours the forms can be removed. Unloading, wheeling and placing 2½ yards of concrete is not a one-person job. It is really a three-person job—two wheeling concrete and one puddling and screeding.

ESTABLISHING THE BUILDING LINE

Fig. 7 shows the concrete-block work started for our addition. The corners should always be laid first when you are starting a wall, but before the block work can start, the building line, the *outside* wall line, must be established. Up to this point we've worked from the *center* line of the footing, with lines stretched from the nails at points A and C (Fig. 1) to saw cuts in batter boards 3 and 4, and with a line stretched from batter board 1 to batter board 2. To

Fig. 7   Laying the concrete-block foundation.

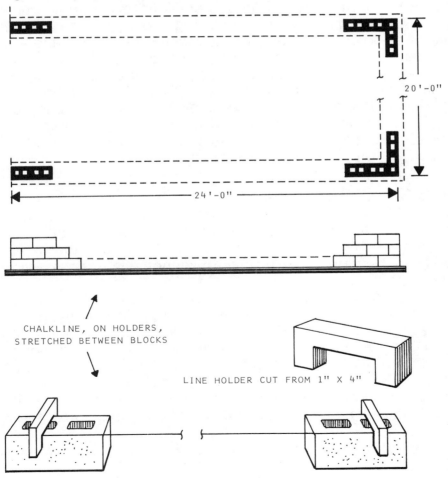

CHALKLINE, ON HOLDERS,
STRETCHED BETWEEN BLOCKS

LINE HOLDER CUT FROM 1" X 4"

establish the building line we need to move all lines *outward* 4 inches from the center of the footing. A new nail should be driven at point A, 4 inches outward from the nail originally used, and at point C, 4 inches outward from the nail originally used, increasing the distance between the nails to 20 feet. (19 feet 4 inches + 4 inches + 4 inches = 20 feet). New saw cuts, 4 inches outward from the ones used to find the center of the footing, must be made at batter boards 3 and 4. Lines stretched from the new nail at point A to the new cut at batter board 3 and from the new nail at C to the new saw cut at batter board 4 will establish the long wall building lines, 20 feet apart. When the new saw cuts are made at batter boards 1 and 2, 4 inches outward from the ones used to find the footing center, and a line is stretched between them, the building line for the short wall—24 feet—will be established (23 feet 8 inches + 4 inches = 24 feet).

A plumb bob hung from the chalk lines will show where lines should be marked on the footing indicating the *outside* edges of the corner and end blocks. When the corner and end blocks have been laid, leveled and plumbed, a chalk line on line holders, shown in Fig. 7, can be used as a guide, eliminating the need to level each individual block as it is laid. This method, used by skilled brick and block layers, will result in a straight, true wall.

When blocks shorter than a full block are needed, they can be cut using a hammer and cold chisel or easily broken at the desired point with a few light taps of a hammer.

LAYING CONCRETE BLOCKS

The first step in any project using concrete blocks is to estimate the number of blocks needed. The actual measurement of a concrete block is 15⅝ inches × 7⅝ inches. When laid, the mortar joints, ⅜ inch, adds up to a total 16 inches × 8 inches laying length.

The simplest method for do-it-yourselfers is to measure the feet, multiply by 12 inches and divide by 16 inches (length of block). A 24-foot wall shown in Fig. 7 would require 18 blocks (24 feet × 12 inches ÷ 16 inches = 18). A 20-foot wall would require 15 blocks (20 × 12 ÷ 16 = 15). Total blocks for one course would be 18 + 15 + 18 = 51 − 1 (½ block will be gained at each corner), or 150 blocks total for three courses. Three corner blocks would be needed at each corner, for a total of 144 common blocks and 6 corner blocks.

A piece of plywood makes a good mortarboard, as shown

in Fig. 8. When mixing mortar, good, clean, sharp sand should be used; the mixture should be 3 parts sand to 1 part brick cement (Kosmortar, Brixment, etc.). Do not use portland cement; brick cements contain lime and should always be used when laying concrete blocks or bricks.

Add only enough water to make a stiff mortar mixture; if the mortar is too wet it will not support the weight of a block. Mortar can be mixed in a wheelbarrow, then placed on a mortar board and sliced off (Fig. 8) as needed. Mortar on a board will tend to dry; keep a can of water on the mortarboard to temper the mortar occasionally. The slices of mortar should be laid on the footing or blocks, as shown in Fig. 9. There is a knack to getting the mortar to

Fig. 8   Cutting slices of mortar to set blocks on.

Fig. 9   Laying slices of mortar to set blocks on.

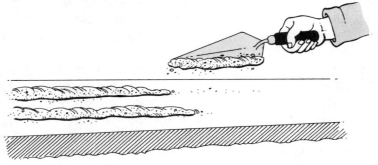

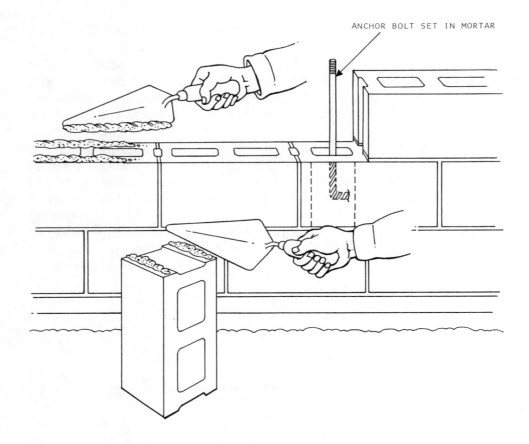

ANCHOR BOLT SET IN MORTAR

Fig. 10   Setting anchor bolts and placing mortar on blocks.

stick to the end of a concrete block, as shown in Fig. 10. The knack is to hit the edge of the block with the edge of the trowel, causing the mortar to slide off the trowel with force and adhere to the end of the block. The block can then be picked up and set into place. Two chalk-line holders, shown in Fig. 7, set in place on blocks at each end of the wall, will hold a chalk line to serve as a guide when laying blocks. A 4-foot level should be used to level the blocks; the trowel can be used to tap the block into place, as shown in Fig. 11. The level should also be used vertically to check that the wall is plumb.

Anchor bolts should be installed at corners and no more than 8 feet apart along the wall line. The anchor bolts will hold the sills (and the building) firmly to the foundation. Anchor bolts should be set in the center of the holes in the block with 3 inches of the thread extending above the top block. After the bolt is set, the hole should be filled with mortar. When setting the bolts, a few minutes spent laying out the floor joist centers, in this case 16 inches, as

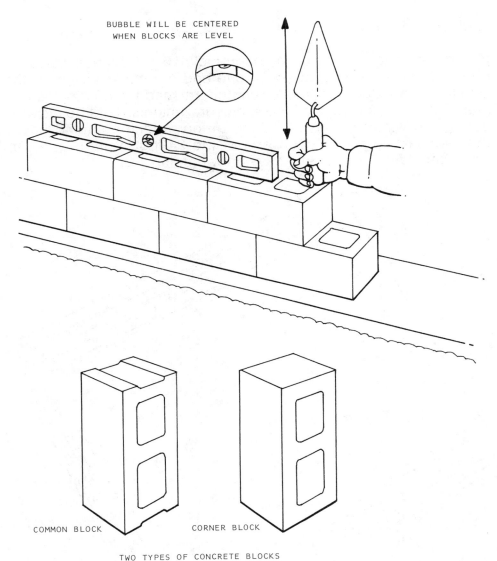

BUBBLE WILL BE CENTERED
WHEN BLOCKS ARE LEVEL

COMMON BLOCK   CORNER BLOCK

TWO TYPES OF CONCRETE BLOCKS

Fig. 11   Using a 4-foot level and trowel to level blocks.

shown in Fig. 14, will prevent the anchor bolts from being directly under joists. After the block laying is finished, the wall should set for 48 hours before the sills are set. Setting the sills is shown in Fig. 12.

FRESH-AIR VENTILATORS

When a building is built on a crawl space, air vents should be placed in the foundation wall. Air circulation is necessary in the warm months to prevent structural damage. Air vents are made to fit into the space occupied by one block, 16 × 8 inches. Special concrete blocks with grooves in one end are made for use with ventilators, and the ventilators can be furnished by the block supplier. This type of vent has a slide closure which should be closed during the winter months. A 20 × 24-foot addition should have four ventilators, one at each end of the 24-foot wall, to provide cross ventilation. A typical vent is shown in Fig. 12.

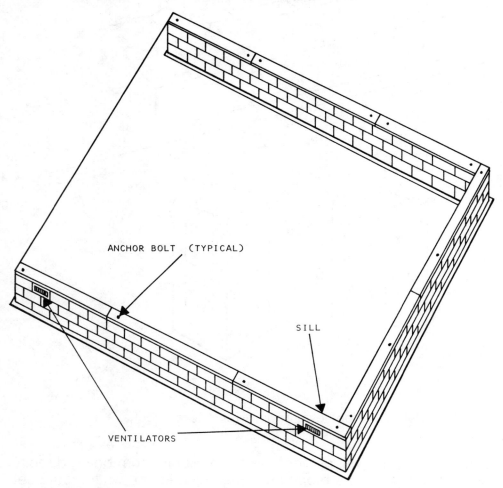

ANCHOR BOLT (TYPICAL)

SILL

VENTILATORS

Fig. 12   Sill is anchored to foundation by anchor bolts, nuts and washers.

plywood has resulted in the increasing use of particle board for subfloors. If particle board is used it should be used in conjunction with plywood. Half-inch CD plywood with a ½-inch particle board applied over the plywood will make a very strong subfloor. The joints of the plywood and the particle board should be staggered. Particle board should *not* be used in a *kitchen* or *bathroom* or in *any* area where it is likely to come into contact with water. Three-quarter-inch CD plywood topped with ¼-inch hardboard (Masonite, etc.) provides a strong smooth surface for vinyl floor covering in kitchens and bathrooms. The subflooring should be protected during construction to prevent water damage.

PLYWOOD

Plywood is cross-laminated (made with several layers each placed at right angles to the adjoining layer), which gives it the property of being able to stand stress in all directions. The outer layers have the grain in the long dimension for greatest strength, and it is recommended that the face grain of plywood be placed across supports in subflooring or roof sheathing, where full strength is important.

HARDBOARD

Hardboard is made from wood fibers and bonded under extreme heat and pressure with a smooth surface completely free from defects. It is available in two basic types: treated, which is highly moisture-resistant, and untreated, for interior use where moisture is not a problem. When used in kitchens and bathrooms, hardboard can be applied with a glue or mastic, making a smooth, nail-free base for floor coverings.

Preventing Moisture Damage

Polyethylene film has proven very successful in preventing moisture damage to wooden structural members when used as a ground cover in crawl spaces. It is available in .004 mil and .006 mil thickness and in rolls from 10 to 20 feet wide and up to 100 feet long.

When more than 1 piece of film must be used, the top sheet should overlap the lower sheet at least 3 feet and be weighted down with bricks or 2 × 4s.

After the floor has been nailed down, the polyethylene film should be applied. It is not necessary to seal it at wall lines, but it should be extended completely to all wall lines and weighted down next to the walls. A good time to install the polyethylene is just before nailing the bridging in the crawl space.

Chapter Four

Framing the Walls

The first step in framing the walls is to mark the stud locations on the plates and shoes. On our original foundation plans the addition is to be 24 feet long and 20 feet wide. The north and south walls when framed would be 24 feet long, the east wall would be 20 feet minus 7 inches, or 19 feet 5 inches. (Minus 7 inches because of the 3½-inch width each of the north and south walls.) The framing drawings show the walls from the inside of the room. The stud and cripple locations should be marked on the plates and shoes, as shown in Fig. 16. When marking use a carpenter's square to make the markings straight and uniform. The X's stand for studs, the C's for cripples. Cripples are additional supports

Fig. 16   Marking locations of cripples and studs on plates and shoe.

which are used where needed in framing and are shown in Figs. 18, 22, 41 and 49. Shoulder studs (Table E) are also called cripples. The wall can be framed laying on the subfloor, then raised up 90 degrees and set into position, as shown in Fig. 17. The shoe can then be nailed through the subfloor into the joists. The shoe should be nailed every 4 feet (or to every third floor joist). The top plate can be added after the wall is raised into position; however, the top plate should overlap at the corners to tie the building together, as shown in Fig. 19. After a wall has been raised into

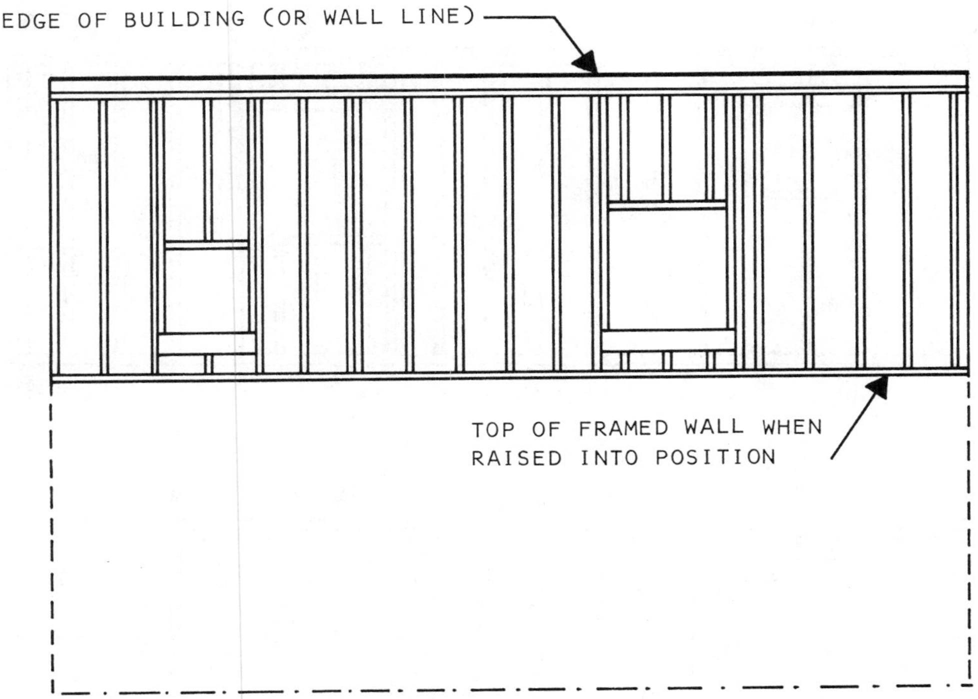

TOP OF FRAMED WALL WHEN
RAISED INTO POSITION

Fig. 17   Framed wall is nailed together on floor, then raised into position at edge of outside wall.

position it should be plumbed and braced, as shown in Fig. 21. Either twelve- or sixteen-penny nails can be used for nailing the framing. As shown in Fig. 17, the shoe is at the *top* of the drawing and very close, within 6 inches of the outside wall. When the wall is raised to a vertical position the shoe will be at the floor and almost exactly in position at the edge of the building. Raising the walls on an addition of this size is normally a three- or four-person job—two or three to raise the wall, the other to first nail the shoe into the joists and then, using a 4-foot carpenter's level, to plumb and brace the wall. The braces should be left in place until the roof trusses have been set.

Wall studs should always be set on 16-inch centers, the only exception being that at corners, doors or windows extra studs may be needed. When studs are set on 16-inch centers a 4-foot, 8-foot or 12-foot sheet of drywall or paneling will fit from the center of one stud to the center of another.

Fire blocks, 2-inch × 4-inch blocks set between studs approximately 48 inches above floor level, are no longer required by most building codes. Check with your local building inspector to find out if fire blocks are required in your area.

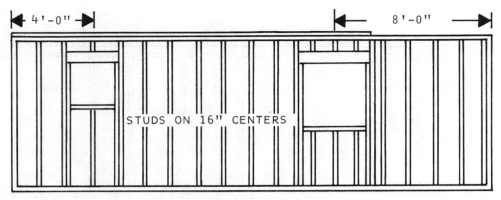

WALL FRAMING–NORTH WALL

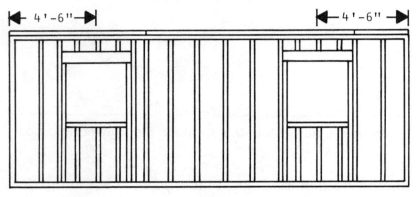

WALL FRAMING–EAST WALL

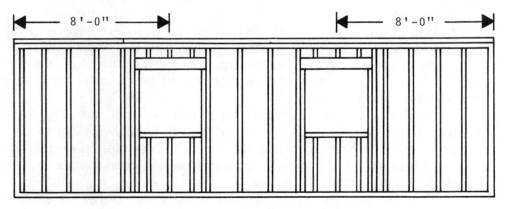

WALL FRAMING–SOUTH WALL

Fig. 18   Details of wall framing.

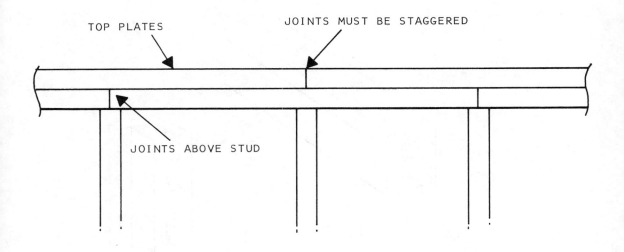

TOP PLATES

JOINTS MUST BE STAGGERED

JOINTS ABOVE STUD

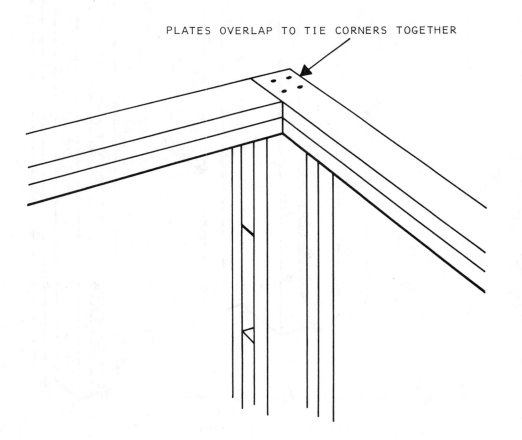

PLATES OVERLAP TO TIE CORNERS TOGETHER

Fig. 19   Details of plate construction.

25

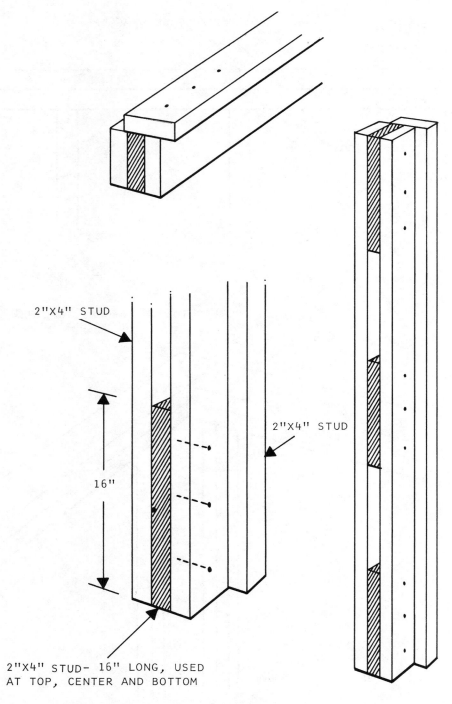

2"X4" STUD

16"

2"X4" STUD

2"X4" STUD- 16" LONG, USED
AT TOP, CENTER AND BOTTOM

Fig. 20   Correct way to build corner posts.

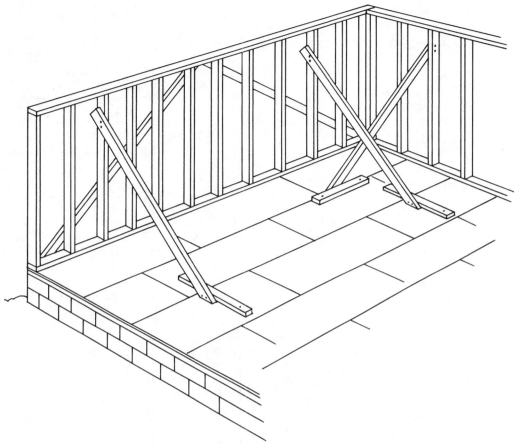

Fig. 21   Walls should be braced after erection.

Framing for doors and windows is shown in Fig. 48 and Fig. 49. Details for building the headers for doors and windows are shown in Table E.

Table D will be helpful in determining the number of studs needed for a wall.

The following table of wall studs will be helpful in determining the number of studs needed for various lengths of walls. In addition to the number shown for straight walls, 2 additional studs will be needed to construct the corner posts at the ends of the wall. Construction of a corner post is shown in Fig. 20. The table is based on studs being placed on 16-inch centers.

TABLE D—WALL STUDS

WALL LENGTH	4'	5'4"	6'8"	8'	9'4"	10'8"	12'	16'	20'	24'	28'	32'	36'	40'	44'
STUDS NEEDED	4	5	6	7	8	9	10	13	16	19	22	25	28	31	34

TABLE E—HEADER SIZES FOR DOORS AND WINDOWS

	SPAN	HEADER SIZE	MADE OF 2″ X's WITH PLYWOOD FILLER
UP TO BUT NOT	2'6″	4″ × 4″	Two 2″ × 4″
MORE THAN	4'0″	4″ × 6″	Two 2″ × 6″
	6'	4″ × 8″	Two 2″ × 8″
	10'	4″ × 10″	Two 2″ × 10″
	12″	4″ × 12″	Two 2″ × 12″

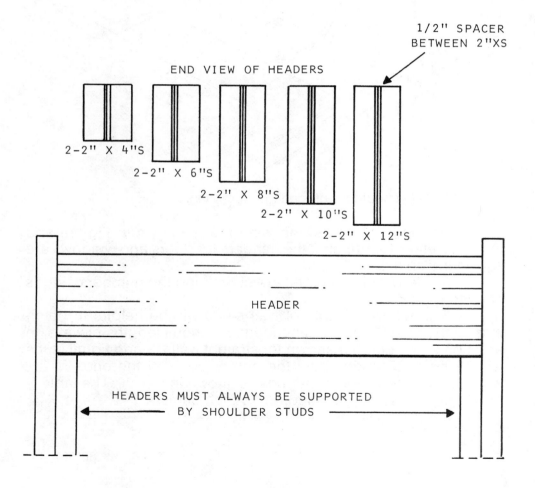

1/2″ SPACER BETWEEN 2″XS

END VIEW OF HEADERS

2-2″ X 4″S

2-2″ X 6″S

2-2″ X 8″S

2-2″ X 10″S

2-2″ X 12″S

HEADER

HEADERS MUST ALWAYS BE SUPPORTED BY SHOULDER STUDS

TABLE E   DETAILS OF HEADERS

Chapter Five

Roof Framing and Roofing

The roof can be framed with roof rafters, but it is much simpler, easier and faster and no more expensive to use prefabricated roof trusses. Your building-material supplier can, when given the width of the building, select the proper type and size trusses for your project. Trusses are designed and fabricated according to the span, the roof pitch and the overhang needed. With a 20-foot span and a 4/12 pitch (the roof slopes 4 inches for every 12 inches in width) it would be standard practice to use a Fink (or W) type truss set on 24-inch centers. Fink-type trusses are used for a roof ending with a gable. Hipped roofs slope down at all four corners from the ridge, thus eliminating the gable end. If a hipped roof is desired your building-material supplier will measure the width of the building, determine the overhang and order the trusses. When they are delivered to the job they should be marked showing their placement because of the assorted sizes and shapes of trusses and other parts of a hip roof.

Our addition uses a Fink truss, shown against the gable end of the existing house in Fig. 22-A. To make the connection between the new roof and the existing building the siding on the existing gable end will have to be cut and removed, as shown in Fig. 22-C. To make this cut in the proper place, set a roof truss on top of the plate and use a short length of 2 × 4 to mark a line 1½ inches above the truss as shown in Fig. 22-B. The siding should then be cut along this line. After the siding is removed set the truss back in place as before and mark a line on each vertical 2 × 4 in the existing gable end. As shown in Fig. 22-D, this line will indicate the top of the 2 × 4s, which, when nailed to the existing gable end, will support the roof sheathing of the addition. Cripples, mitered to the angle of the supporting 2 × 4s and nailed to the existing gable-end framing, provide additional supporting strength.

There are several advantages in using prefabricated roof trusses: they require no load-bearing interior walls for support, they are uniform in size and because of their design they can be placed on 24-inch centers.

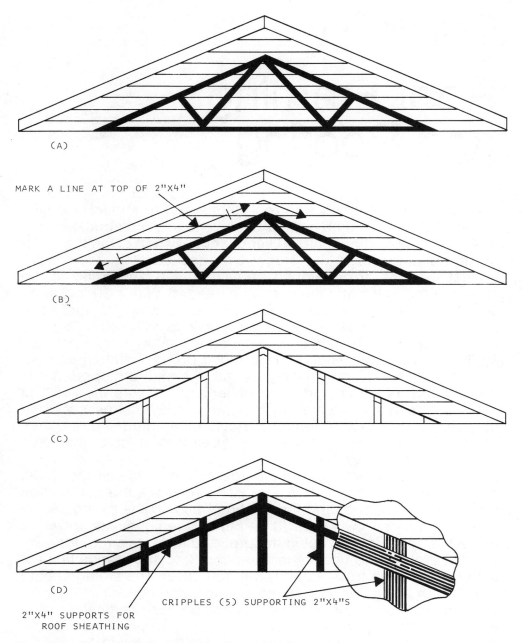

(A)

MARK A LINE AT TOP OF 2"X4"

(B)

(C)

(D)

2"X4" SUPPORTS FOR
ROOF SHEATHING

CRIPPLES (5) SUPPORTING 2"X4"S

Fig. 22   Construction details at existing gable end.

When the trusses are set in place, as shown in Fig. 23, they should be toenailed into the top plate. Use a level to check that the trusses are plumb (vertically straight), then nail a 1 × 4 from truss to truss, near the tops of the trusses, to hold them in place until the sheathing is applied.

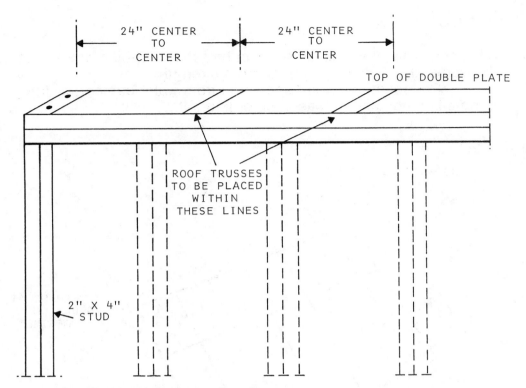

Fig. 23   Marking locations of roof trusses.

TABLE F—ROOF TRUSSES															
BUILDING LENGTH	16′	18′	20′	22′	24′	26′	28′	30′	32′	34′	36′	38′	40′	42′	44′
TRUSSES NEEDED	9	10	11	12	13	14	15	16	17	18	19	20	21	22	23

Table F is based on roof trusses being placed on 24-inch centers, and on the use of Fink-type trusses. If an overhang with soffit is desired, an extra truss will be needed.

SHEATHING THE ROOF

CD-grade plywood is the most commonly used type of roof sheathing. It is available in ½-inch, ⅝-inch and ¾-inch thicknesses. Although the ½-inch thickness is adequate, ⅝-inch plywood will add rigidity to a roof, especially when roof trusses are set on 24-inch centers. When the trusses are set to 24-inch centers the 4-foot × 8-foot sheets of plywood can be applied with a minimum of cutting. The joints should be staggered for strength, as shown in Fig. 24, and eight-penny galvanized nails should be used at 6-inch intervals. Either full or half sheets should be used at the gable-end overhang for maximum strength.

Protect the plywood sheathing from water damage by covering it immediately with asphalt roofing felt.

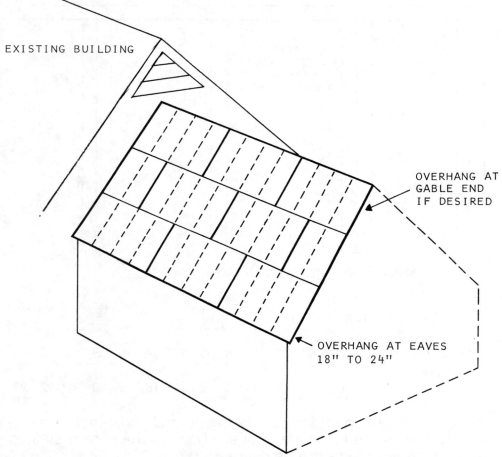

EXISTING BUILDING

OVERHANG AT
GABLE END
IF DESIRED

OVERHANG AT EAVES
18" TO 24"

Fig. 24   Joints in plywood roof sheathing should be staggered as shown and nailed at 6-inch intervals.

FLASHINGS

After nailing down the sheathing next to the existing wall, the flashing, shown in Fig. 25, should be inserted underneath the siding, as shown in Fig. 26. Tin snips can be used to cut the flashing, as shown in Fig. 25-B, to make it fit at the ridge. The flashing will not have to be nailed where it extends up and under the siding. Nail to the roof sheathing only, with nails close to the edge of the flashing where the roofing will cover them.

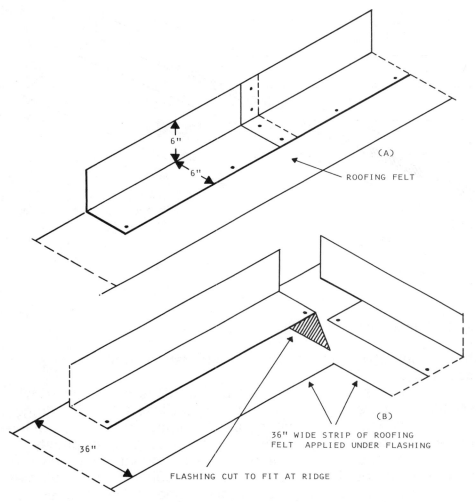

6"

6"

(A)

ROOFING FELT

(B)

36" WIDE STRIP OF ROOFING
FELT APPLIED UNDER FLASHING

36"

FLASHING CUT TO FIT AT RIDGE

Fig. 25   Flashings needed at connection to existing building.

The flashing should be made of copper or 24-gauge galvanized steel, formed to an L shape and 6 inches × 6 inches, as
shown in Fig. 25.

Fibered asphalt roofing cement should be applied over
the nail heads and at the point where the existing siding meets the
flashing, as shown in Fig. 26, to seal the connection between the
new roof and the existing building.

ROOFING

There is a trend among roofing manufacturers to discontinue making asphalt roofing shingles. They are switching to fiber-glass
shingles. Asphalt shingles are made of asphalt-saturated felt coated on both sides with waterproof asphalt and mineral granules.

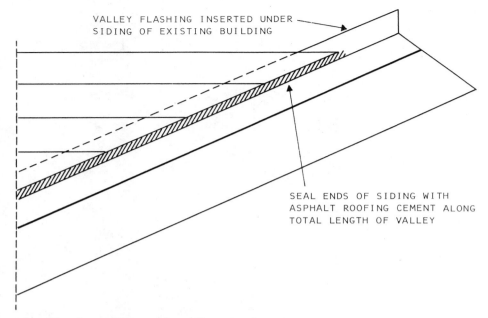

VALLEY FLASHING INSERTED UNDER
SIDING OF EXISTING BUILDING

SEAL ENDS OF SIDING WITH
ASPHALT ROOFING CEMENT ALONG
TOTAL LENGTH OF VALLEY

Fig. 26   Installing and sealing the flashing.

Asphalt shingles carry a Class C fire rating from Underwriters Laboratory.

Fiber-glass shingles have a fiber-glass base coated with asphalt and mineral granules and carry a Class A UL fire rating.

Both asphalt and fiber-glass shingles are made in different weights; in either type heavier weight means better quality.

When roofing an addition, you should match the new roof to the existing roof in both tab shape and color. If spare shingles were left from the existing construction, one of the shingles can serve as a pattern both for tab shape and color. If no spares were left, take a piece of cardboard to the roof, slip it under a shingle, as shown in Fig. 27, and trace around the tabs. Building-material suppliers should be able to match the pattern using either the shingle or the cardboard template and furnish color samples.

The first step in roofing a new house or addition is to apply roofing felt over the sheathing. Generally, 15-pound felt is used; a roll of 15-pound felt contains 432 square feet. The length of the roof times the width (one side) times 2 equals the total number of square feet needed. There will be some waste, so allow for this when ordering the felt. Roofing felt should be overlapped at least 3 inches, as shown in Fig. 29; wood lath temporarily nailed at intervals of 6 or 8 feet will aid in securing the felt until the shingles are applied.

Roofing shingles are sold in "squares" (100 square feet) and are packaged in bundles, three bundles to a square. To esti-

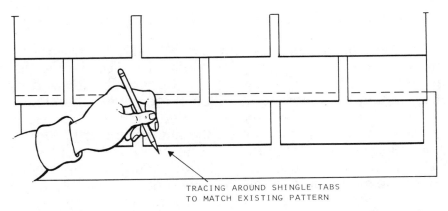

TRACING AROUND SHINGLE TABS
TO MATCH EXISTING PATTERN

Fig. 27   How to make roofing shingle pattern.

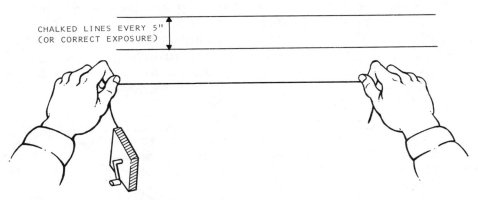

CHALKED LINES EVERY 5"
(OR CORRECT EXPOSURE)

Fig. 28   Using a chalk line to mark correct exposure.

mate the amount of shingles needed, measure from the eave to the ridge, multiply by the length of the roof, double this figure (to allow for both sides of the roof) and divide by 100. The result will be the number of squares needed.

Example: A roof is 24 feet long and 12 feet from eave to ridge. How many squares of roofing will be needed?

```
   24      288         5.76
  ×12      ×2    100)576.00       or    5¾ squares
   48      576        500
   24               760
  288             
                    600
                    600
```

We will need 5¾ squares, plus the caps at the ridge, shown in Fig. 33, so we would order 6 squares.

One-inch galvanized roofing nails should be used and should be driven at right angles to the roof to avoid damaging the shingles. Metal starter strip, shown in Fig. 30, should be used to stiffen the starting edges of the shingles at the eaves. The starter strip should overhang the edges of the sheathing or the fascia board ¾ inch. An easy way to line up the starter strip is shown in Fig. 30. A chalk line is stretched from one end of the roof to the other, ¾ inch out from the edge of the sheathing or the fascia board. Hold the starter strip to this line when nailing it in place.

The correct steps for applying the roofing shingles are shown in Fig. 31. After the starter strip is nailed on (A), the first row of shingles is nailed down; note that this row is reversed (upside down). The nails should be near the top of the shingle, where they will be covered by the first row of regular shingles. After the first (reversed) row is nailed down, the first regular row is nailed directly on top of the reversed row, as shown in Fig. 31-C. The portion of the shingle which shows, the *exposure*, is governed by the pattern and type of shingle. The supplier should furnish you with the

Fig. 29   Applying 15-pound felt underlayment.

15 LB. FELT UNDERLAYMENT

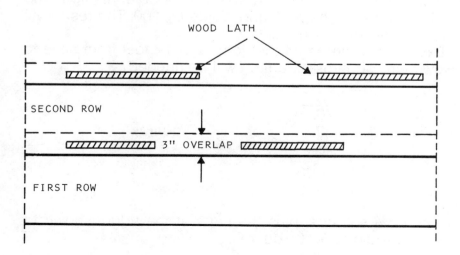

WOOD LATH

SECOND ROW

3" OVERLAP

FIRST ROW

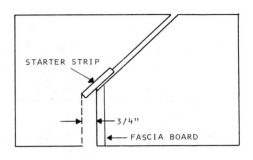

STARTER STRIP

3/4"

FASCIA BOARD

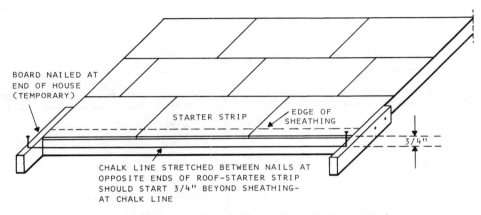

BOARD NAILED AT
END OF HOUSE
(TEMPORARY)

STARTER STRIP

EDGE OF
SHEATHING

3/4"

CHALK LINE STRETCHED BETWEEN NAILS AT
OPPOSITE ENDS OF ROOF-STARTER STRIP
SHOULD START 3/4" BEYOND SHEATHING-
AT CHALK LINE

Fig. 30   Using a chalk line as guide to install starter strip.

recommended exposure when the shingles are purchased. When matching an old roof, if the shingles are a perfect match with the existing roof, measure the exposure of the existing shingles and duplicate this on the new roof. Assuming the exposure to be 5 inches (more or less standard), use a chalk box and line and snap chalk lines every 5 inches above the top of the first row of shingles, as shown in Figs. 28 and 31-D. Continue laying the shingles until you reach the ridge; on the first side to reach the ridge, stop the shingles at the ridge line, cutting off some of the top shingle if necessary. When the other side is brought to the ridge line it should be turned over the ridge not more than 2 inches and nailed down.

The cap row is made by cutting shingles into three pieces, as shown in Fig. 33, bending them over the ridge and nailing them down. The top cap row should be embedded in fibered asphalt roofing cement to prevent any possibility of leaks. Cover any exposed nails with a dab of roofing cement.

Where the roofing meets the flashing at the edge of the existing building, the shingles should be embedded in roofing cement, as shown in Fig. 32.

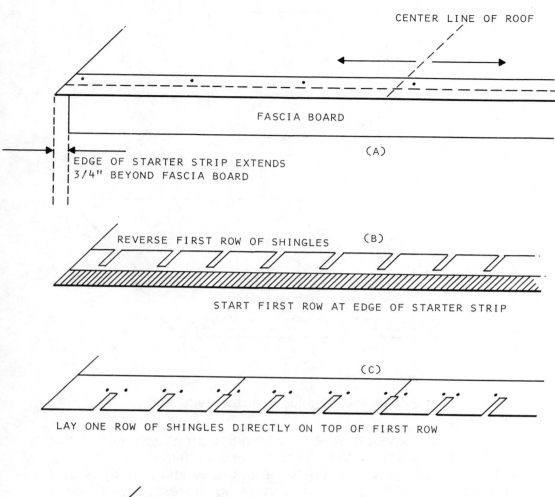

CENTER LINE OF ROOF

FASCIA BOARD

(A)

EDGE OF STARTER STRIP EXTENDS
3/4" BEYOND FASCIA BOARD

REVERSE FIRST ROW OF SHINGLES (B)

START FIRST ROW AT EDGE OF STARTER STRIP

(C)

LAY ONE ROW OF SHINGLES DIRECTLY ON TOP OF FIRST ROW

(D)

CHALKED LINES ON
ROOFING FELT

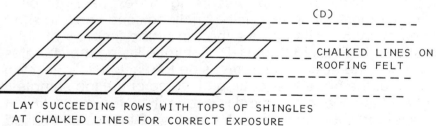

LAY SUCCEEDING ROWS WITH TOPS OF SHINGLES
AT CHALKED LINES FOR CORRECT EXPOSURE

Fig. 31   Steps in correct application of roofing shingles.

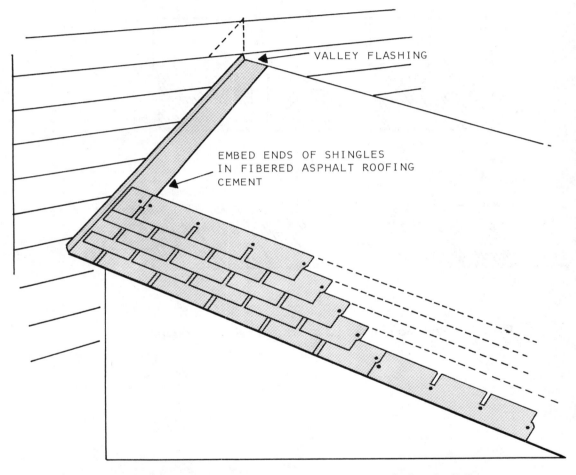

VALLEY FLASHING

EMBED ENDS OF SHINGLES
IN FIBERED ASPHALT ROOFING
CEMENT

Fig. 32   Sealing roofing shingles at connection to existing building.

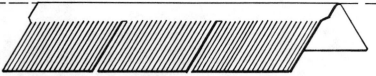

RIDGE LINE OF ROOF

TOP ROW ON ONE SIDE TURNED OVER RIDGE AND NAILED DOWN

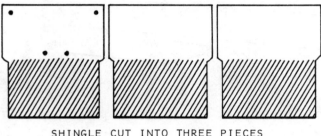

SHINGLE CUT INTO THREE PIECES

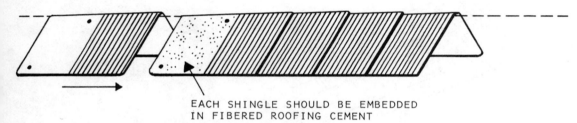

EACH SHINGLE SHOULD BE EMBEDDED
IN FIBERED ROOFING CEMENT

Fig. 33   Roofing is finished at ridge as shown above.

Chapter Six

Enclosing the Building: Sheathing and Siding

WALL SHEATHING

Strength, insulation values and ease of application should all be considered before you decide which type of sheathing to apply to the outside walls. CD plywood in either ½-inch or ⅝-inch thickness will add strength but have little insulation value. Asphalt-impregnated fiberboard is often used, although it has a low insulation rating. There are several brands of energy-saving fiberboard, approximately ⅛-inch thick and specially treated, with reflective foil on both sides and with U-values of approximately .2. Thicknesses 1 inch and up of Styrofoam and Fiberglas High-R Sheathing have high insulation values. Your building-material supplier can furnish you with information on the relative values of these and other types of sheathings. He will also have complete application instructions for the particular type of sheathing you select. If a wood sheathing is used, a layer of 15-pound asphalt felt should be applied to the sheathing before the siding is nailed on.

SIDING

When you are building an addition, your first thought on siding might be to match the original building. This may be possible, but if the original building is of brick or stone it will involve considerable labor toothing in the new brick or stone, and the new materials will not blend into the older building for several years. Wood, hardboard or aluminum siding should not be too hard to match in size, and if wood or hardboard siding is used, the first time the original building and the addition are painted the addition will blend in. If aluminum siding is used on the existing house and also on the addition, the new siding may take several years to weather and match the existing siding. Why worry about matching the existing siding? Let the new addition contrast with the existing building; just because it contrasts does not mean it conflicts.

Building-material suppliers stock a wide variety of lap sidings—aluminum, cedar, hardboard and plywood, to name a few. Some types of sidings are prefinished; others require staining or painting.

When a lap siding is applied the bottoms of the boards must be level. The easiest way to ensure this is to use a chalk line as a guide.

Example: Using a 9-inch wide hardboard siding, the top of the first course should be 8 inches above the bottom of the foundation sill. The bottom of the course will then be 1 inch below the bottom of the sill. Drive a nail, leaving 1 inch exposed, 8 inches above the bottom of the sill at each end of the building. Stretch a chalk line or staging line between the nails. Pull the line as *tight* as possible and secure it. Hang a line level on the line; if the level indicates that one end of the line is high, lower the high nail until the line is level. Starting flush with one corner, tack the first course of boards to the sheathing, holding the tops of the boards even with the chalk line. Unless the siding is being nailed to wood sheathing the butt joints should always center on a wall stud. Do not drive the nails in fully at this time. When the first course has been applied stand back and inspect it visually. If the first course is not level, make any necessary adjustments, then finish nailing. Some types of sidings in long lengths will have a tendency to sag in the middle; holding the top of the board to the chalk line when nailing will ensure a straight course. After the first course has been applied drive a nail 8 inches above the first nail (at each end), stretch the line again and follow the same procedure for the second and succeeding courses. In this example we used a 9-inch siding; if siding of another width is used, change the distance between the nails (8 inches) to fit the different width.

Using a line level as a guide is shown in Fig. 34.

PLYWOOD SIDING

Plywood panel sidings, in 4-foot × 8-foot sheets, are made with waterproof glues and are available in a great variety of finishes and colors. Plywood siding is easy to work with, can be cut to fit around windows and doors and is made with ship-lap edges for perfect joints. When carefully applied the joints are almost impossible to detect. Batten boards 1 inch × 3 inches or 1 inch × 4 inches should be used at corners, and if used at the bottom edges

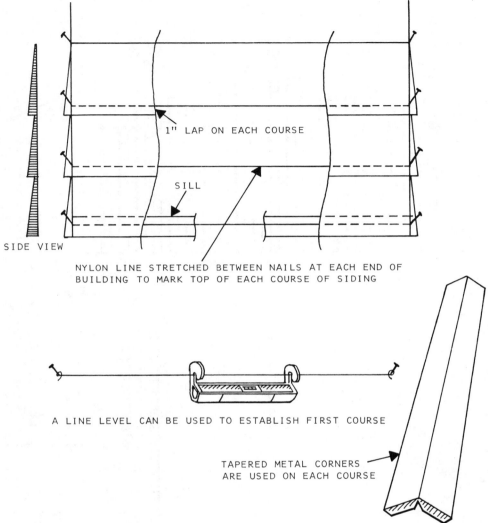

SIDE VIEW

1" LAP ON EACH COURSE

SILL

NYLON LINE STRETCHED BETWEEN NAILS AT EACH END OF
BUILDING TO MARK TOP OF EACH COURSE OF SIDING

A LINE LEVEL CAN BE USED TO ESTABLISH FIRST COURSE

TAPERED METAL CORNERS
ARE USED ON EACH COURSE

Fig. 34   Steps in application of lap siding.

of the siding will improve the overall appearance. When plywood
siding is used on walls over 8 feet in height, the top sheets or cut
pieces of siding should overlap the lower sheets. This is done by
nailing ⅝-inch furring strips (if ⅝-inch siding is used) over the
sheathing, as shown in Fig. 35. To avoid rust staining, aluminum or
galvanized nails should be used to nail the siding.

Plywood sidings have grooved or textured finishes, the
grooves must be vertically true or plumb when the sheets are
applied. A carpenter's level, 48 inches long, should be used to
plumb the edges of the siding as it is applied.

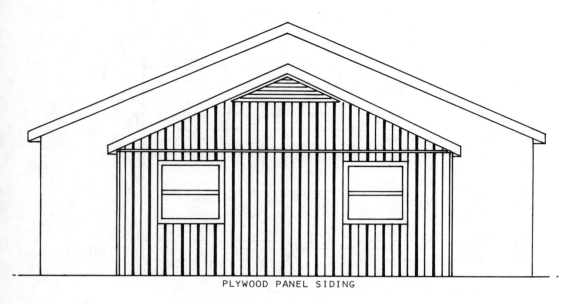

PLYWOOD PANEL SIDING

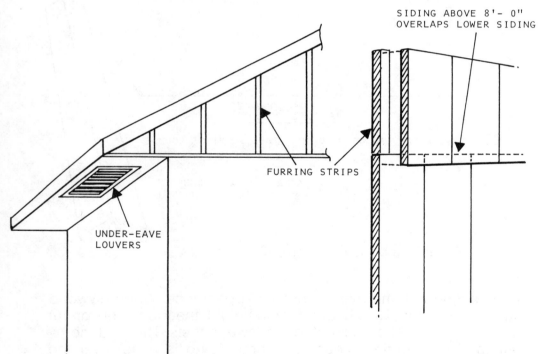

SIDING ABOVE 8'- 0"
OVERLAPS LOWER SIDING

FURRING STRIPS

UNDER-EAVE
LOUVERS

Fig. 35   Details of siding and eaves.

SOFFITS AND LOUVERS

The soffit is applied to the bottom side or overhang of the roof trusses or rafters. The soffit seals the opening between the top plate of the outside wall and the trusses or rafters, not against the flow of air but to prevent birds, squirrels, etc., from gaining entrance to the above-ceiling space or attic. Air flow is necessary for proper ventilation of this space. The soffit can be made of plywood or cement board and if made of these materials louvers should be installed as shown in Fig. 35 and Fig. 36. The number of louvers needed will depend on the size and type selected and the size of the space to be ventilated. There are disadvantages in this type of soffit: it is not the easiest to install and it should be painted periodically.

Prefinished aluminum ventilated soffit is the preferred type to use; it installs easily, is made in 12 foot-long panels and can be cut with tin snips, it is perforated for air circulation and it will not rust or require painting for many years.

Turbine ventilators, ridge ventilators or wall louvers can be used to allow the air coming in through the soffit to circulate; the wall louver shown in Fig. 36 is the easiest to install and the most practical, trouble-free type.

Fig. 36   Louvers are mounted in soffit board to allow air circulation in attic space.

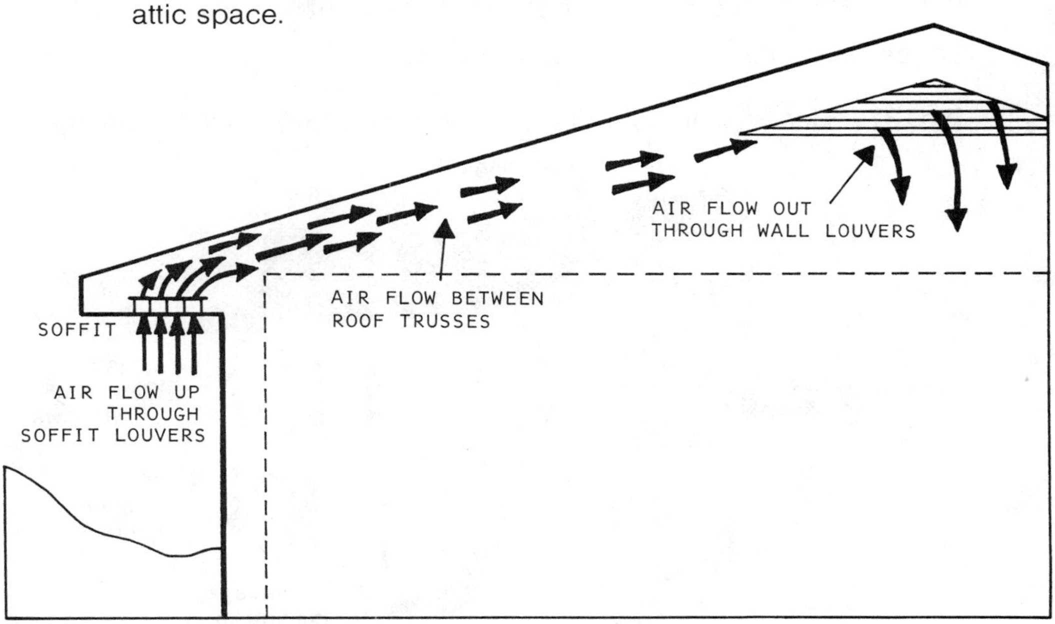

SOFFIT

AIR FLOW UP
THROUGH
SOFFIT LOUVERS

AIR FLOW BETWEEN
ROOF TRUSSES

AIR FLOW OUT
THROUGH WALL LOUVERS

Chapter Seven

Insulation

The high cost of energy used for both heating and cooling demands that the best types and methods of insulation be used in home construction. The American Society of Heating, Refrigerating and Air-Conditioning Engineers, Inc. has determined that heat loss through a typical conventionally insulated home is as follows:

> Through basement floors 1 percent
> Through basement walls 20 percent
> Through basement frame walls 17 percent
> Through basement cracks in windows/doors/walls
>   38 percent
> Through basement windows 16 percent
> Through basement doors 3 percent
> Through basement ceilings 5 percent

These heat losses are shown in Fig. 37.

Fig. 37   Heat loss through a typical conventionally insulated home.

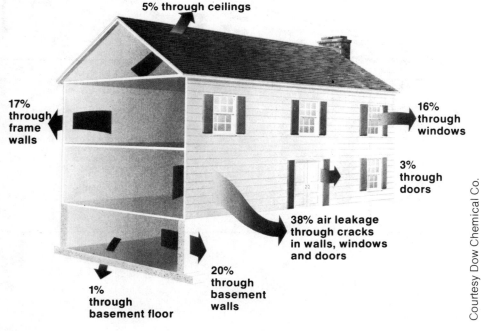

5% through ceilings

17% through frame walls

16% through windows

3% through doors

38% air leakage through cracks in walls, windows and doors

1% through basement floor

20% through basement walls

Courtesy Dow Chemical Co.

The two common terms used in energy codes are R-values and U-values. Don't let this confuse you: either can be converted to the other by dividing it into 1.

Adequate insulation may require an insulation board on the outside of the stud walls plus batt-type insulation between the studs. This is especially true in frame construction using 2-inch × 4-inch-stud walls because the 2 × 4s will actually measure 3½ inches. One way to achieve an R-19 rating for a 2 × 4 wall cavity is to use:

	R-value
Aluminum siding	.61
1-inch Styrofoam T.G.	5.41
R-13 batt insulation	13.00
½-inch gypsum drywall	.45
Outside air film	.17
Inside air film	.68
Total R-value	20.32
U-value	.05

Fig. 38 Corner bracing recessed into studding will permit the application of Styrofoam insulation over the corner areas of the house.

Courtesy Dow Chemical Co.

Corner bracing, shown in Fig. 38, is required for all nonstructural sheathings; hardboard, foamboard or fiberboard. When the corner bracing is recessed into the studding, full-thickness Styrofoam can be applied to the corners as well as the rest of the house. The Fiberglas wall insulation is shown in Fig. 39. The batting is made in 16-inch widths to fit between the studding, and the asphalted kraft paper has flanges at the edges for easy stapling installation. R-13 insulation can be used with 3½-inch studding. The insulation should fit tightly to both the double plate at the top and the shoe plate at the bottom of the wall. Care should be taken to see that the flanges do not gap and permit vapor penetration. Cut the insulation 1 inch longer at the top and bottom of wall cavity; the facing can then be peeled back and stapled to the top and bottom

Courtesy Owens-Corning Fiberglas Corporation

Fig. 39   Applying kraft-faced insulation in wall cavity.

plates. If the facing is torn, it should be resealed to remain an effective vapor barrier. Insulation of ceilings will add greatly to the comfort of a room by eliminating downdrafts. In Fig. 40, we see a workman installing ceiling insulation. The insulation should be pushed up through the joist spaces, then pulled back down even

Fig. 40   Installing Fiberglas batting insulation between ceiling joists.

with the bottom of the ceiling joist. This assures full R-value performance by preventing compression of the insulation. When High-R Batts are used, this method allows the batts to expand over the top of the joist. Overlap the top plate with the insulation, but if eave vents are used, do not block the vents. The ends of insulation should butt tightly against each other. The facing is flanged and should be stapled to the joists. Obstructions such as electrical boxes for light fixtures should have insulation installed over them, with this exception: *Caution: Insulation must be kept at least 3 (three) inches away from recessed light fixtures.* Fig. 41 shows the minimum R-values recommended for various sections of the country.

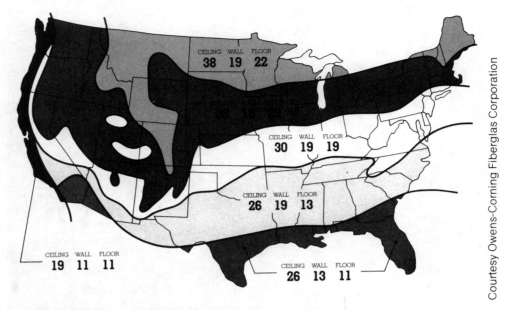

Courtesy Owens-Corning Fiberglas Corporation

Fig. 41   This map shows the minimum insulation R-values recommended by Owens-Corning for various parts of the country.

Chapter Eight

Inside Framing: Partition Walls, Doors and Windows

INTERIOR NONLOAD-BEARING WALLS

Part of the excitement and satisfaction of remodeling a home is the creation of new living space, finishing a basement or attic area, converting a garage into a family room or dividing one large room into several more efficient spaces. All of these projects call for building and framing nonbearing partitioned walls. Because a nonbearing wall carries no weight from above, its construction is simpler than a load-bearing wall. It requires a top plate, a sole plate and studs set on 16- or 24-inch centers.

   The first step is to mark off on the floor the location of the wall, as shown in Fig. 42. When the measurement has been made

Fig. 42   Marking the wall location with a chalk line.

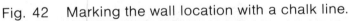

Courtesy Georgia-Pacific Corp.

and the chalk line is snapped, the sole plate can be cut and nailed to the subfloor or existing floor. If the top plate runs at a 90-degree angle to the ceiling joists, the plate can be cut and nailed directly to the joists. If the top plate runs parallel to the ceiling joists, the framing can be as shown in Fig. 43. Two-by-fours are cut and nailed into place between the floor joist, the top plate is then nailed to the 2 × 4s and the studs can then be measured to length between the top plate and the sole plate.

If the room has no ceiling, the ceiling joists will be visible, but if the room has a ceiling, after the location of the top plate has been marked, small holes can be made above the plate location and the ceiling joists can be located.

Fig. 43   Details of wall framing when wall is parallel with joists or ceiling rafters.

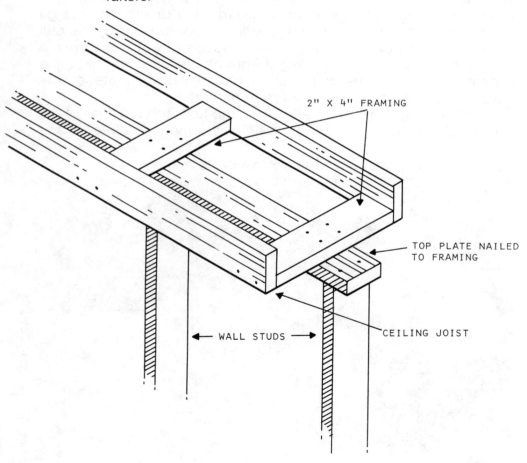

2" X 4" FRAMING

TOP PLATE NAILED TO FRAMING

WALL STUDS

CEILING JOIST

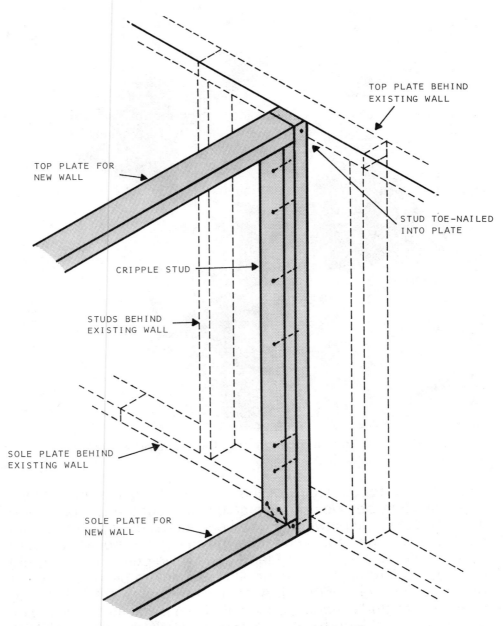

TOP PLATE BEHIND
EXISTING WALL

TOP PLATE FOR
NEW WALL

STUD TOE-NAILED
INTO PLATE

CRIPPLE STUD

STUDS BEHIND
EXISTING WALL

SOLE PLATE BEHIND
EXISTING WALL

SOLE PLATE FOR
NEW WALL

Fig. 44   How to frame new wall against existing wall.

When installing a partition wall in an existing room, with walls and ceilings already in, you could run into problems. The wall line of the new partition may be at a point between the studs of the intersecting end wall. If this is the case, the end stud should be cut the full length between the existing ceiling and the floor. It can then be nailed into the hidden top plate and the hidden sole plate of the existing wall, as shown in Fig. 44. If the same condition

Fig. 45   Checking to make certain that the first stud is plumb.

exists at the opposite end of the wall, do the same thing there. Then measure between the two studs, and cut the top plate and the sole plate to fit between the two end studs. Both plates should be a tight fit. If the floor is concrete, the sole plate should be nailed with concrete nails; if a wooden floor, twelve-penny nails should be used. Before nailing the end studs, check to be sure they are plumb, as shown in Fig. 45. After the two plates are nailed, the intermediate studs should be toenailed into place as shown in Fig. 46, on either 16-inch or 24-inch centers. Nailing the sole plate is shown in Fig. 47.

FRAMING DOORS AND WINDOWS
The size and type of doors and windows will determine the size of the rough openings for the doors and windows. Prehung doors and preassembled windows are available and save time and money for do-it-yourselfers by providing a fast, easy and secure method of installation.

Fig. 46   Studs are toenailed into plate and shoe.

Fig. 47   Concrete nails should be used when nailing to
a concrete floor.

Regardless of whether you decide on prehung or standard doors and windows, you will need to know the dimensions of the rough opening required. This information should be furnished with the units.

The size of the double headers, shown in Fig. 48 and Fig. 49, for the doors and windows will depend on the width of the units. Two-by-sixes can be used for spans up to 4 feet and 2 × 8s for spans up to 6 feet. For wider spans up to 8 feet, sliding glass doors or French doors, 2 × 10s will provide adequate support. The header is a lintel carrying the weight from above, and this framing member should not be undersized. Note that, as shown in Table E, the header is made of 2 pieces of 2 × 6, 2 × 8, etc., with a ½-inch spacer between them.

Fig. 48   Details for framing a door opening.

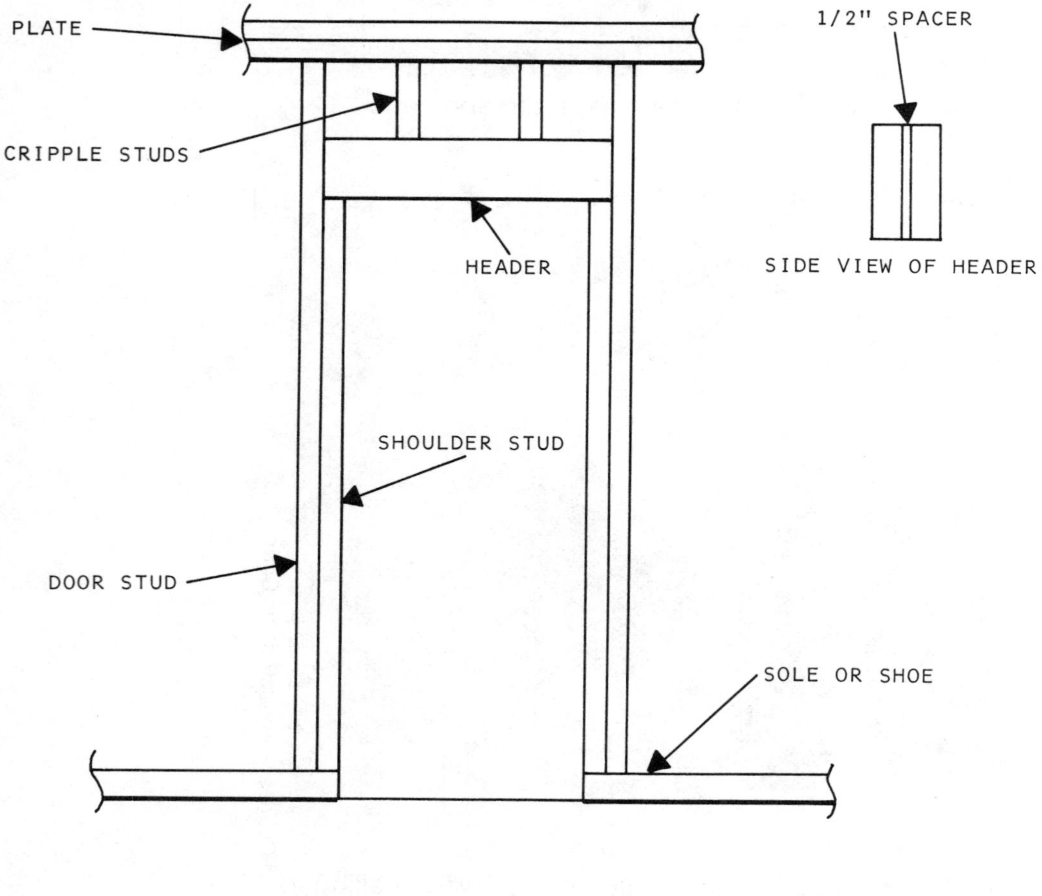

PLATE

CRIPPLE STUDS

HEADER

1/2" SPACER

SIDE VIEW OF HEADER

SHOULDER STUD

DOOR STUD

SOLE OR SHOE

Installing Prehung Doors and Preassembled Windows
Prehung doors and preassembled (set-up) windows make the task of installing these items a simple one for the do-it-yourselfer.

Windows
Windows can be purchased prime-painted or vinyl-clad, if desired, to further lessen the work of finishing. They are available with different types of casings. The side and top members, which hold the window sash in place, are called jambs. When selecting the windows for a project, make sure the jamb depth is equal to the thickness of the wall.

Fig. 49   Framing details for a window opening.

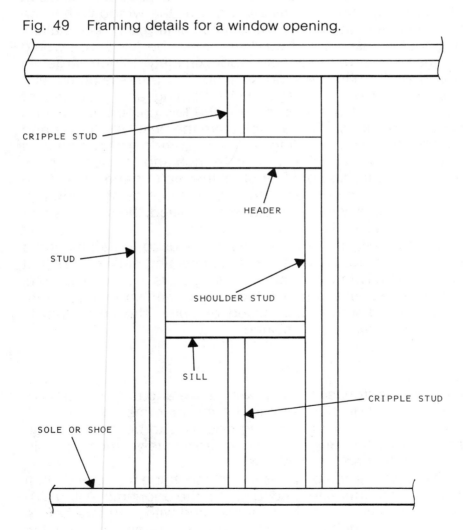

Example: If the wall is a 2 × 4 wall with 1-inch foam insulation board on the outside and ⅝-inch drywall on the inside, a total of 3½ inches + 1 inch + ⅝ inch, the jamb must be 5⅛ inches in depth. This is necessary in order for the inside casing to be nailed to the edge of the jamb without leaving a gap between the inside wall and the casing. If the exterior wall—brick, stucco, aluminum or wood siding—requires a deeper jamb, the window supplier can advise you when the windows are ordered. Roofing felt should be fitted into the rough framing before the windows are set to help prevent air or water leakage; here also, the suppliers can advise you what will be needed for the type of window you select. A wood or metal drip cap must be installed above the window after the window is in place and before the outside siding, etc., is installed. Knowing that preassembled windows would be used, you would have framed the rough opening approximately 1 inch larger in height and width than the window jamb. The preassembled window should be set into place in the framing and plumbed or leveled, using wooden shims to raise the low corners or take up the space at the sides. Shim shingles are the best type to use and are available from your building supplier in random width bundles of from 3 inches to 12 inches and in 14-inch and 16-inch lengths. They are tapered from about ⅝ inch at the top to paper-thin at the bottom and should be driven in from both sides as shown in Fig. 50. After the jamb is wedged in place and nailed, the shingles can be broken or sawed off even with the jamb.

Casing nails, if available, should be used to nail the jamb to the casing (casing nails are similar to finishing nails except that they have a slightly larger head). If casing nails are not available, finishing nails are satisfactory. Either type of nail should be countersunk and filled with plastic wood or putty. The finish trim is installed after the walls are finished.

Doors

Prehung doors, purchased set up with jambs and hinges installed, can be easily and quickly set into the rough frame and secured. Knowing in advance that a prehung door was to be installed, you would have made the rough framing approximately 1 inch larger in height and width than the door casing.

The casing should be set into place in the frame and shim shingles inserted where needed to plumb the doorjamb, using a 4-foot level. The casing must also be checked with a square before nailing the jamb to the frame; shim shingles are wedges, and if too

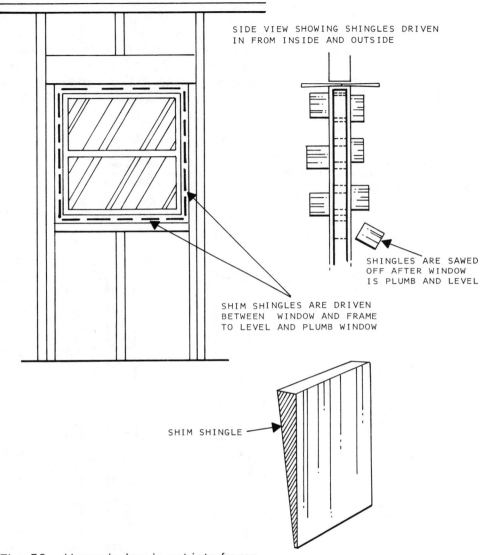

SIDE VIEW SHOWING SHINGLES DRIVEN
IN FROM INSIDE AND OUTSIDE

SHINGLES ARE SAWED
OFF AFTER WINDOW
IS PLUMB AND LEVEL

SHIM SHINGLES ARE DRIVEN
BETWEEN WINDOW AND FRAME
TO LEVEL AND PLUMB WINDOW

SHIM SHINGLE

Fig. 50   How window is set into frame.

much force is applied through the shims, the casing can be forced out of square. If the casing is forced out of square and then nailed into the frame, the casing will be damaged if the nails have to be pulled. Check carefully that the casing is plumb and square before nailing.

Roofing felt should be applied to the framing before setting the prehung unit into place. The amount of felt needed will depend on the door selected and the building supplier can advise you on how much felt should be used.

The shims should be set close together when the casing is plumbed and the casing nails, set at 12-inch intervals, should be driven through the shims. A final word of caution: be sure to check the corners of the casing before nailing to make certain that the shims have not pushed the casing out of square. Fig. 51 shows how the prehung door is set into the frame.

Fig. 51   How pre-hung door is set into frame.

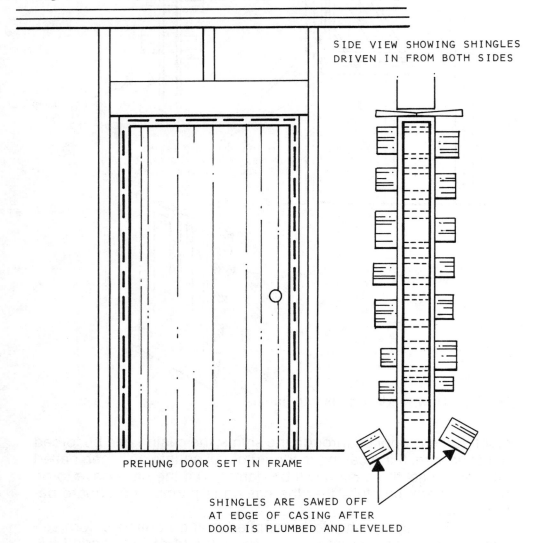

SIDE VIEW SHOWING SHINGLES DRIVEN IN FROM BOTH SIDES

PREHUNG DOOR SET IN FRAME

SHINGLES ARE SAWED OFF AT EDGE OF CASING AFTER DOOR IS PLUMBED AND LEVELED

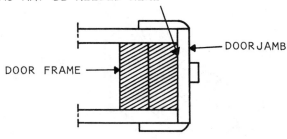

SHIMS MAY BE NEEDED HERE

DOOR FRAME

DOOR JAMB

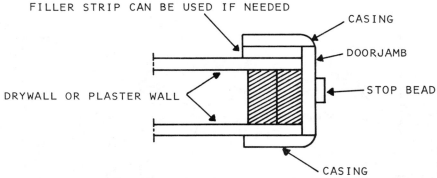

FILLER STRIP CAN BE USED IF NEEDED

CASING

DOORJAMB

DRYWALL OR PLASTER WALL

STOP BEAD

CASING

Fig. 52   Top view of door framing and casing.

HANGING STANDARD DOORS

Doors are hung either right or left hand; this is determined by facing the door from the "outside." With an inside (passage) door the outside is the side with the hinges not visible and the door opening away from the viewer. Standing on the outside, if the hinges are on the left side it is a left-hand door, if they are on the right it is a right-hand door.

The fitting of the door hinges to the door and the frame can be done in either of these ways:

1. using a router and template;
2. using a hammer and wood chisel.

If the router-and-template method is used, instructions will be furnished with the templates. A butt gauge, shown in Fig. 53, is a useful tool for marking the areas to be cut out. If many doors are to be hung it would pay to buy one, but doors can be hung accurately without using a butt gauge.

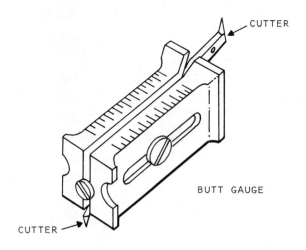

CUTTER

BUTT GAUGE

CUTTER

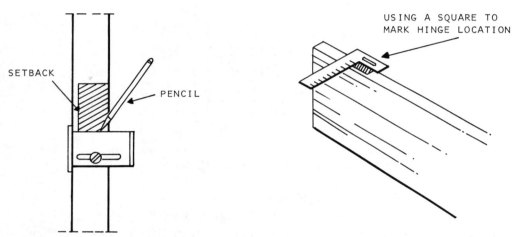

SETBACK

PENCIL

USING A SQUARE TO
MARK HINGE LOCATION

Fig. 53   Tools used to mark hinge location.

There are two types of hinges commonly used on doors, swaged and not swaged. As shown in Fig. 54, when closed the swaged hinge comes together all along its length, while the hinge which is not swaged leaves a gap at the hinge-pin side. A device on one end of the butt gauge measures the offset for swaged hinges. What size hinges should you use when hanging a door? Common practice is to use two 3½-inch hinges on inside doors and three 4-inch hinges on outside (entrance) doors. Entrance doors are generally heavier and therefore require more and larger hinges. The center of the hinge on an interior door should be about 9 inches from the top, the center of the lower hinge at 13 inches from the bottom; on entrance doors the same measurements are used with the center hinge centered on the other two, as shown in Fig. 54.

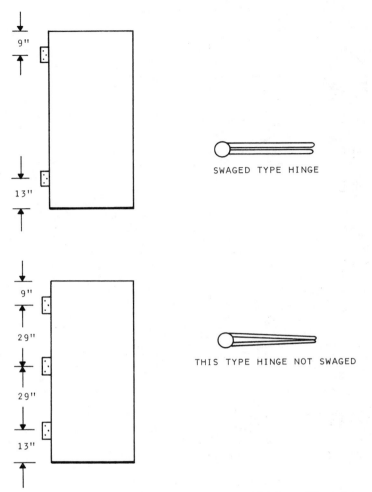

Fig. 54 Standard measurements for installing hinges on a 2 feet 6 inches × 6 feet 8 inches door.

If you use the butt gauge, set the butt gauge flange against the side of the door and mark the hinge location. Then measure the width of the hinge and set the butt gauge for this width. Again, set the butt gauge against the door and mark the hinge width. Note that about ¼ inch will be left at the side. This is called the setback; it is not cut out and will hide the hinge when the hinge is mounted. After the door is marked set the butt gauge against the jamb and mark the same measurement on the jamb. Use a sharp wood chisel to outline the marked area, then carefully cut out the wood to a depth equal to the thickness of the hinge. When the wood is cut to the correct depth the surface of the hinge will be flush with the wood surface.

Hold the hinge in place on the door and the jamb and use

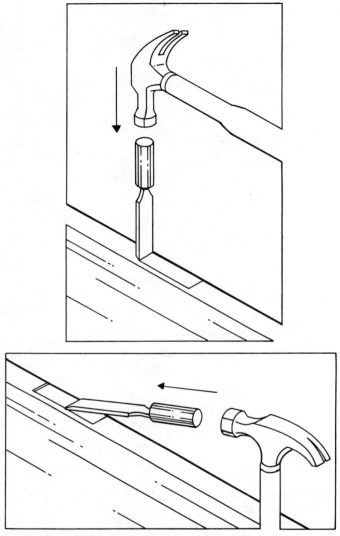

Fig. 55   Using a hammer and chisel to mark and
cut out wood to install hinges.

a nail or a carpenter's awl to make a small hole to start the screws, then tighten the hinge in place. Set the door into the opening and insert the hinge pins. If the door does not swing freely the hinges may need adjusting. If necessary, plane off any high spots on the door.

INSTALLING LOCKSETS

After you have hung the door, the next step is to install the lock-sets. Passage knob sets are used for doors which do not lock. Privacy locksets locked by a turn button are usually used on bedroom and bathroom doors. Entrance doors use key locksets, and these are made in several different types, with and without deadbolts. Installation instructions and templates are furnished with each type of lockset. Read the manufacturer's instructions carefully before starting work: a mistake could damage, even ruin, an expensive door. The standard height for locksets is 38 inches above the floor to the center of the knob; separate deadbolt locks are usually mounted 6 inches above a standard lockset.

THRESHOLDS

Thresholds are used at entrance doors to provide a tight seal at the bottom of the door. A different type of threshold, usually just a metal strip, is used where floors of different types or heights meet, as between a carpeted bedroom floor and a tiled bathroom floor. Entrance-door thresholds are usually installed after the door is hung. There are two common types: the bumper type, which is installed on the outside of the door, and the standard threshold,

Fig. 56   Two commonly used types of threshold.

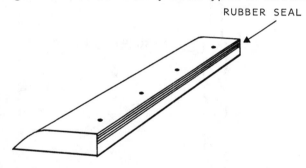

RUBBER SEAL

BUMPER THRESHOLD

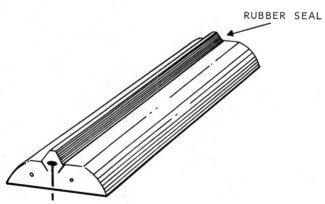

RUBBER SEAL

STANDARD THRESHOLD

which has a rubber sealer which presses against the bottom of the door when the door is closed. If the bumper type is used the threshold should be cut to fit between the sides of the frame; then the door should be closed and the threshold set against the door on the outside. Using the screw holes in the threshold as a guide, mark the screw holes on the floor. An awl or a nail should be used to start the screw holes in a wood floor; if the floor is concrete an electric drill motor and a carbide tipped drill can be used to drill holes for the plastic anchors as shown in Fig. 57. Quarter-inch plastic anchors should be used, with a ¼-inch masonry (carbide-tipped) drill bit, the holes should be drilled about 1 inch deep and the loose dust blown out, after which the plastic anchor can be pushed into the hole and the threshold set in place and fastened down to the anchors with metal screws. Either wood screws or metal screws can be used to secure the threshold to the floor.

Fig. 57   Setting anchors in a concrete floor.

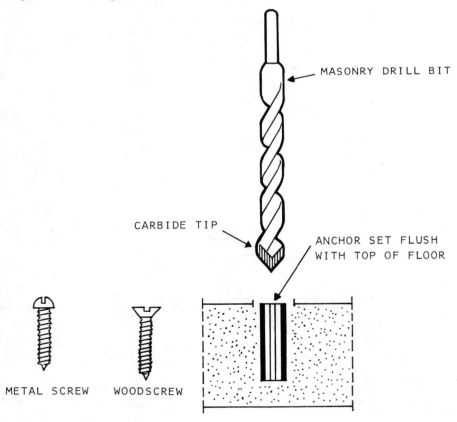

MASONRY DRILL BIT

CARBIDE TIP

ANCHOR SET FLUSH
WITH TOP OF FLOOR

METAL SCREW   WOODSCREW

Chapter Nine

Plumbing the Addition

WHY ROUGH-IN MEASUREMENTS ARE IMPORTANT

If you plan to remodel a bathroom, adding ceramic tile to the present wall, the tile will add approximately ⅜ inch (tile plus mastic). The minimum measurement, from the center of the outlet of the toilet to the finished wall line, should not be less than 10 inches. This is shown in Fig. 58. The measurement from the center of the

Fig. 58   Rough-in measurements for a toilet.

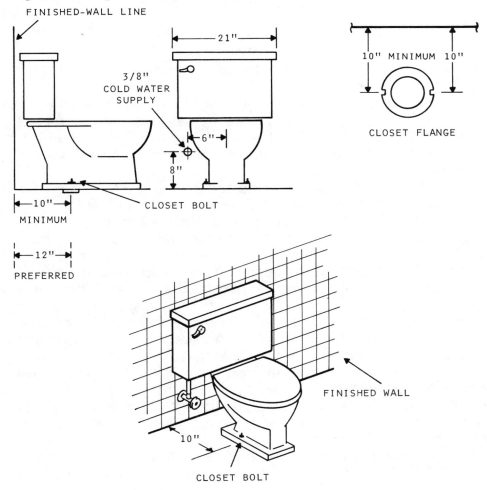

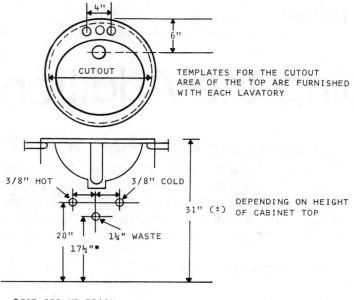

TEMPLATES FOR THE CUTOUT
AREA OF THE TOP ARE FURNISHED
WITH EACH LAVATORY

*FOR POP-UP DRAIN;
19" FOR CHAIN AND PLUG DRAIN.
CHAIN AND PLUG DRAIN RARELY USED IN RESIDENCES.

Fig. 59   Rough-in measurements for a self-rimming lavatory.

closet bolts (which is also the center of the bowl outlet and the center of the closet flange) to the wall is called the rough-in measurement. Toilets are made in 10-inch, 12-inch and 14-inch rough-in measurements. (Ten-inch and 14-inch rough-in toilets are *not* made by *all* manufacturers.)

If a plumber is called to replace a toilet or a toilet bowl, the first thing he will do is measure from the wall to the center of the closet bolts. He will then order a 10-inch, 12-inch or 14-inch rough-in unit, depending on the measurement. A bowl or closet combination (tank and bowl) made for a 10-inch rough-in can be set on a 14-inch roughed-in opening, but it would leave an approximately 5-inch space between the back of the tank and the wall. A 12-inch or 14-inch rough-in unit could not be set on a 10-inch roughed-in opening. A ½-inch plus-or-minus variation is acceptable. As an example, a 12-inch rough-in toilet should fit an opening roughed in at 11½ inches or at 12½ inches.

The water supply for the two piece toilet (tank and bowl) should be roughed in at approximately 8 inches above the finished floor and 6 inches off center (to the left, as you face it) of the toilet. If a one-piece toilet is used, the manufacturer's roughing-in sheet should be followed.

The rough-in measurements for a lavatory are shown in Fig. 59. This is a vitreous china self-rimming lavatory set in the cabinet shown in Fig. 60. If the roughing in is done correctly, final connections are quick and easy. If you are planning an addition or a remodel involving the installation of new plumbing fixtures, when selecting the new fixtures, ask for the rough-in measurements for the fixtures to ensure the proper locations of water and waste piping.

Fig. 60   Finished bathroom showing self-rimming lavatory mounted in cabinet top.

ROUGHING IN THE PLUMBING

It will be necessary to make a new connection into the existing sanitary drainage piping and extend this connection to the new half bath shown in Fig. 61. The easiest way to make this connection is to remove a cleanout plug, screw a threaded adapter, called an MIP (male iron pipe) adapter, into the fitting and install new piping to the point of use. Cleanouts are not always available: it may be necessary to call in your local plumber and have him or her provide a connection. Some possible points of connection for the new piping are shown in Fig. 62: (A) is a horizontal connection into an existing soil-pipe line extending through a wall; (B) is a connection into a test tee; (C) is a connection into a base ell (elbow). The method of connection would be the same for any of these fittings. First the plugs must be removed. If they are the original plugs they will probably be made of brass and they will have corroded, making them impossible to unscrew. They will have to be cut out, using a cold chisel and a hammer. This is not

Fig. 61   Rough-in measurements for half bath.

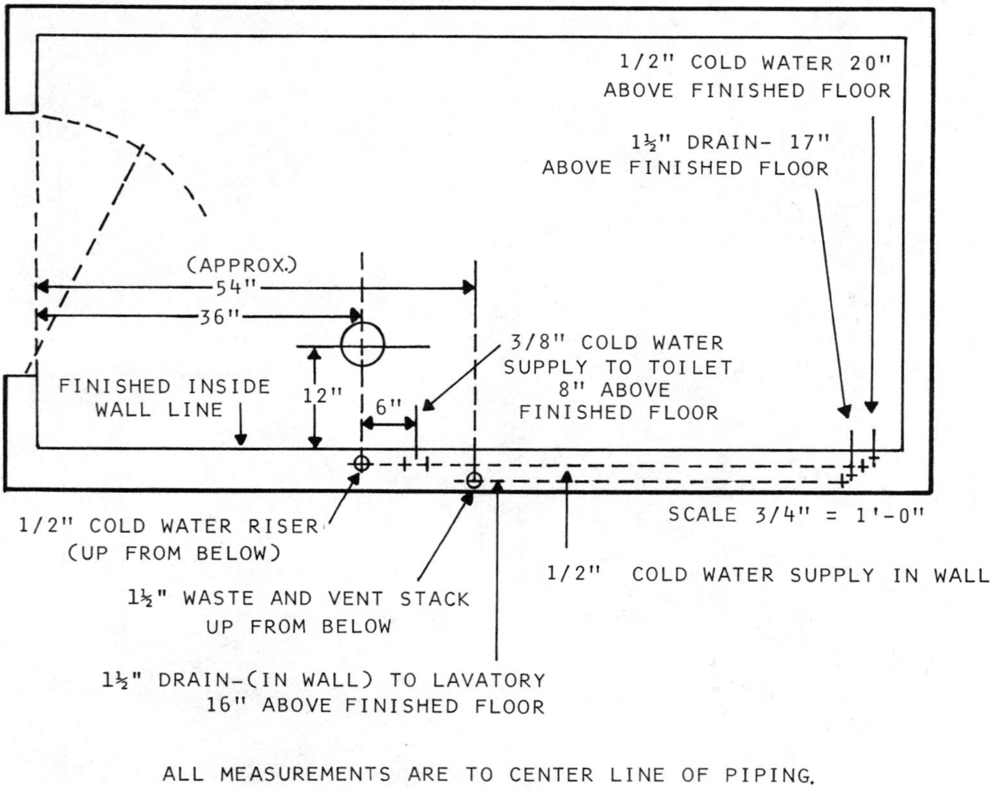

1/2" COLD WATER 20"
ABOVE FINISHED FLOOR

1½" DRAIN- 17"
ABOVE FINISHED FLOOR

(APPROX.)
54"

36"

3/8" COLD WATER
SUPPLY TO TOILET
8" ABOVE
FINISHED FLOOR

FINISHED INSIDE
WALL LINE

12"   6"

1/2" COLD WATER RISER
(UP FROM BELOW)

1½" WASTE AND VENT STACK
UP FROM BELOW

SCALE 3/4" = 1'-0"

1/2"   COLD WATER SUPPLY IN WALL

1½" DRAIN-(IN WALL) TO LAVATORY
16" ABOVE FINISHED FLOOR

ALL MEASUREMENTS ARE TO CENTER LINE OF PIPING.

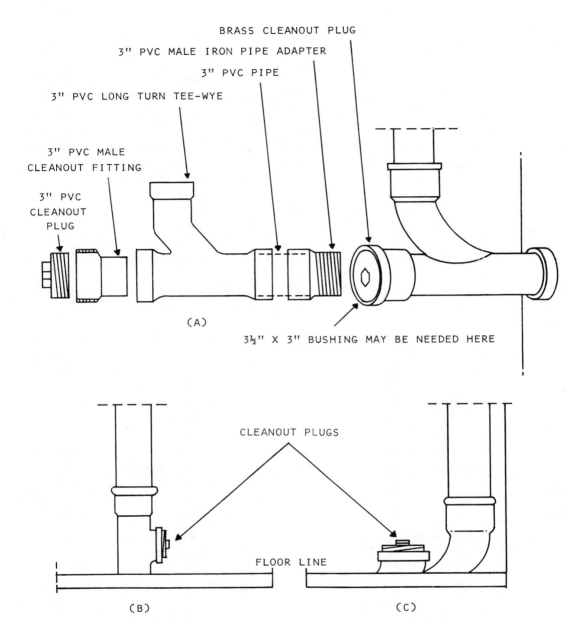

BRASS CLEANOUT PLUG

3" PVC MALE IRON PIPE ADAPTER

3" PVC PIPE

3" PVC LONG TURN TEE-WYE

3" PVC MALE
CLEANOUT FITTING

3" PVC
CLEANOUT
PLUG

(A)

3½" X 3" BUSHING MAY BE NEEDED HERE

CLEANOUT PLUGS

FLOOR LINE

(B)   (C)

Fig. 62   Possible locations for connecting into existing soil pipe.

difficult—brass plugs are less than ⅛ inch in thickness—but be careful using the chisel not to damage the threads in the fitting. A cleanout plug measuring 3½ inches across will be for 3-inch ID (inside diameter) piping. Plumbing piping is always measured by inside diameter. There are times, however, when you must measure the outside diameter in order to know the inside. A cleanout plug which has a 4-inch outside diameter is a 3½-inch plug. Three-and-a-half-inch pipe is not used for plumbing drainage, so a

3½-inch × 3-inch bushing would be needed. If the plug measures 4½ inches, a 4-inch × 3-inch bushing would be needed. A bushing is a pipe fitting which is inserted into the hub or female opening of another pipe fitting to reduce the opening or size of the female fitting. *EXAMPLE:* If a 3-inch opening in a pipe is available and it is desired to extend a 2-inch pipe from the available opening to another point, a bushing (3-inch × 2-inch) could be used to make the transition. Pipe bushings are made of:

    A. copper, for soldered joints
    B. steel or cast iron, with pipe threads on the outside or male end of the bushing, and pipe threads on the inside or female end of the bushing, for insertion into a threaded fitting.
    C. PVC (poly-vinyl-chloride) or ABS (acrylonitrile-butadiene-styrene) for use with PVC or ABS piping.

Plastic pipe is now widely used for both new and repair work and can be easily installed by the do-it-yourselfer. Plastic pipe and fittings are made in two types: PVC and ABS. The pipe and fittings are cemented together with special cements. PVC pipe and PVC fittings should be cemented with PVC cement; ABS cement should be used on ABS pipe and ABS fittings. If for some reason the pipe and fittings must be mixed, use an all-purpose cement made for either type of plastic. For best results use all PVC or all ABS if possible. From this point on in explaining roughing-in methods, the term PVC should be taken to mean PVC or ABS. Drainage fittings should be marked PVC-DWV, and drainage pipe should be marked Schedule 40. PVC-DWV fittings are made with a recessed socket to provide a smooth, unobstructed passageway for waste water and solids. Pressure fittings, if used, would leave a shoulder of the pipe exposed, causing an obstruction. The pipe can be cut with a hacksaw or a cross-cut handsaw, and care should always be taken to cut the end square, using a miter box if one is available. When joints are properly made, they will fit snugly, as shown in the inset in Fig. 63.

    When cementing PVC pipe the proper method is to apply cement to the socket of the fitting (the brush will be in the cap of the can of cement), then to the spigot end of the pipe, and *immediately* fit the two together. When the fitting must be turned in a certain direction, be sure it is turned in this direction when the pipe and fitting are assembled; you have at the most 2 *seconds* to turn it; after that the cement will have set sufficiently to make any adjustment impossible.

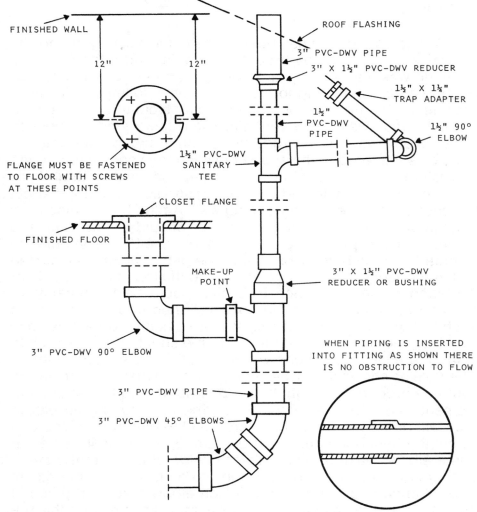

FINISHED WALL

12"   12"

FLANGE MUST BE FASTENED
TO FLOOR WITH SCREWS
AT THESE POINTS

FINISHED FLOOR

CLOSET FLANGE

1½" PVC-DWV
SANITARY
TEE

MAKE-UP
POINT

3" PVC-DWV 90° ELBOW

3" PVC-DWV PIPE

3" PVC-DWV 45° ELBOWS

ROOF FLASHING

3" PVC-DWV PIPE

3" X 1½" PVC-DWV REDUCER

1½" X 1¼"
TRAP ADAPTER

1½"
PVC-DWV
PIPE

1½" 90°
ELBOW

3" X 1½" PVC-DWV
REDUCER OR BUSHING

WHEN PIPING IS INSERTED
INTO FITTING AS SHOWN THERE
IS NO OBSTRUCTION TO FLOW

Fig. 63   Drainage fittings needed to rough in half bath.

The enlarged floor plan of the half bath in Fig. 61 shows that we have to cut two holes through the floor—one to provide for the water closet (toilet) connection, the other for the drain piping to the lavatory. Fig. 63 shows the piping and fittings needed for the half bath. The main soil and waste piping is 3-inch. (Soil piping is the term for piping carrying the discharge from a water closet.) The vent pipe for the water closet is in part also the drain piping for the lavatory. This is called a "wet" vent and is permissible under the Uniform Plumbing Code. Above the tee into which the toilet drainage is connected, the piping can be reduced to 1½-inch.

Whenever it is necessary to remove a cleanout to provide a new opening, as shown in Fig. 62, another cleanout should be added as shown.

After a connection has been made to the existing piping, the new waste piping must be brought to a point under the new bath where the waste and vent stack will turn up through the partition wall, and after connecting to the water closet (toilet) and lavatory, continue out through the roof. Holes must be cut through the floor for the 1½-inch waste and vent stack and the 3-inch connection for the toilet. From the enlarged floor plan in Fig. 61, we see that the toilet is centered in the space between the vanity cabinet and the existing building wall. In order to keep the lavatory drain fairly short, the 1½-inch waste and vent piping should be installed about 60 inches from the existing building wall. A plumb bob should be hung through the hole and the 3-inch pipe should be turned up and centered on the plumb bob. A short piece of pipe should be cemented into the bottom, or outlet, side of the 3-inch sanitary tee and the reducer should be cemented into the top of the tee. A 3- or 4-foot-long piece of 1½-inch PVC pipe could be temporarily inserted through the hole in the floor and then into the top of the 3-inch × 1½-inch reducer. It should not be cemented at this time. The 3-inch tee with the short piece of pipe should be inserted into the 45-degree ell but not cemented. With the closet flange setting in the hole as shown in Fig. 64, a plumb bob centered in the flange should be dropped through the flange and an end-to-center measurement taken for the 3-inch PVC pipe and 90-degree elbow, which will be turned up and centered on the plumb bob. You can cement the 3-inch tee and short piece of 3-inch pipe into the 45-degree elbow, making certain that the side opening of the tee is pointing squarely at the plumb bob hanging through the closet flange. You can now cement the 3-inch elbow and short piece of pipe into the side opening of the tee, making sure that the 90-degree elbow is turned straight up. With the closet flange in place, a measurement can be made from the end of the hub or makeup point of the flange to the makeup point of the 90-degree elbow, and this piece of pipe can be cut and cemented into the top of the 90-degree elbow. This flange should set on the *finished* floor, and since ⅛-inch-thick vinyl floor tile will be used on top of the subfloor, ⅛-inch shims should be set under the flange temporarily. The flange can be cemented to the 3-inch riser, and care must be taken to ensure that the recesses for the closet bolts are at right angles to the stud wall, as shown in Fig. 63. After the flange has been cemented, the pipe will be held rigidly in place. The 1½-inch pipe placed temporarily into the 1½-inch hub can be removed and a measurement taken up to the center of the 1½-inch sanitary tee. This will be an end-to-center (of the tee) measurement, as shown in

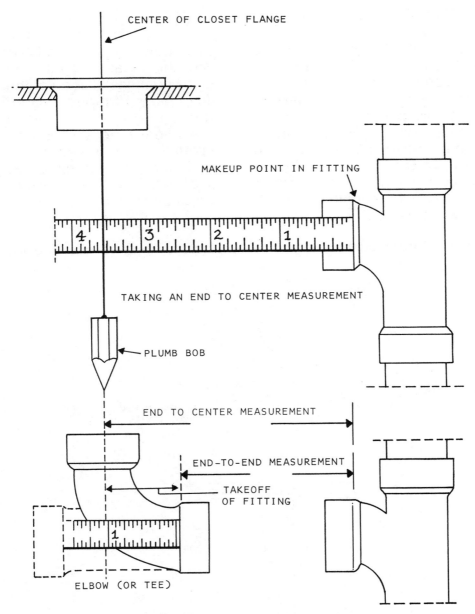

CENTER OF CLOSET FLANGE

MAKEUP POINT IN FITTING

TAKING AN END TO CENTER MEASUREMENT

PLUMB BOB

END TO CENTER MEASUREMENT

END-TO-END MEASUREMENT

TAKEOFF OF FITTING

ELBOW (OR TEE)

Fig. 64   How to make end-to-center measurements.

Fig. 64. The arm or piece of 1½-inch pipe extending from the tee to the 90-degree elbow, where the piping comes out of the partition wall under the vanity top, will be approximately 3 feet long. The sanitary tee has a ¼-inch pitch, or slope, built into the branch, or side, opening. The drainage piping should rough in at 17 inches, which means that the center of the tee should be ¾ inch lower to allow for the ¼-inch pitch, or 16¼ inches above the finished floor. (The ⅛ inch for vinyl tile is negligible for this measurement.) The partition studs will have to be notched to permit installation of the arm; the 90-degree elbow and an 18-inch piece of pipe will complete this part of the roughing-in. From the top of the 1½-inch sanitary tee, the vertical pipe becomes a vent pipe and can be extended upward through the double plate of the partition and, if local codes require it, increased to 3 inches and extended through the roof. Care should be taken when setting the roof flashing to be certain that the top extends under the shingles, as shown in Fig. 65.

Fig. 65   Above-floor waste and vent piping showing extension through roof.

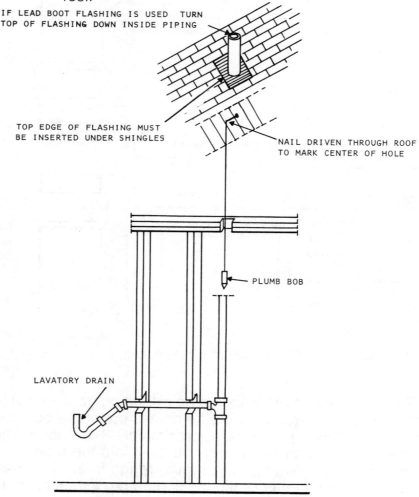

A connection to the cold-water piping must be made in the existing building, as shown in Fig. 66. This connection must be extended into the crawl space under the addition and up through the partition wall. The piping need only be ½-inch ID copper tube (⅝-inch OD). Fig. 67 shows the piping running through the partition wall and stubbed out at the toilet location and into the cabinet under the lavatory. You may have noticed that we made a cold-water connection only in the existing building and extended a cold-water connection only to the half bath. This is for a very good reason: to save energy by a *point-of-use* water heater. In a conventional plumbing installation for a room addition, both hot- and cold-water piping would have been extended into the half bath.

Fig. 66   Making new connections to galvanized pipe or copper tubing.

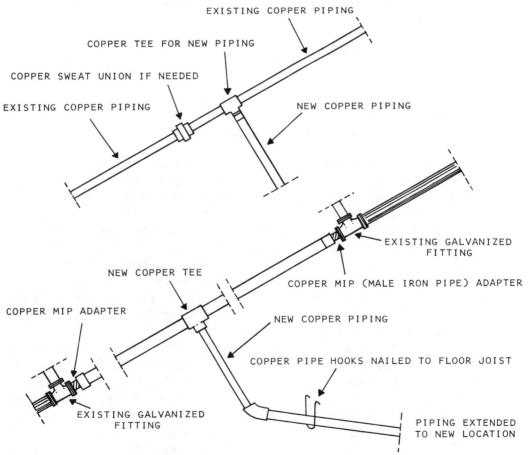

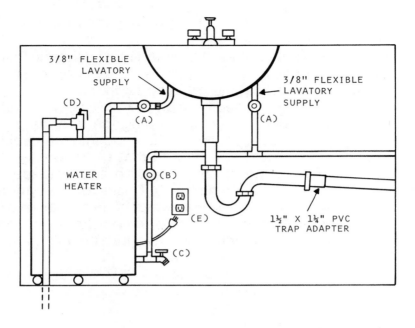

(A) 1/2" COPPER X 3/8" OD (OUTSIDE DIAMETER) COMPRESSION STOP (VALVE)
(B) 1/2" COPPER SWEAT STOP (VALVE)
(C) 1/2" BOILER DRAIN VALVE
(D) 3/4" TEMPERATURE AND PRESSURE RELIEF VALVE
(E)   110 VOLT RECEPTACLE

Fig. 67   Water heater installed in cabinet.

When hot water was needed at the lavatory, it would have been necessary to let the water run, wasting water, until the cold water in the hot-water piping was replaced by the incoming hot water. (Hot water lying dormant in the hot-water piping will quickly cool.) This is a condition which I know is all too common in most homes today and there are ways to overcome it: (1) a point-of-use water heater and (2) a circulating water line. The circulating hot-water system is explained elsewhere in this book, and the point-of-use system is explained in this way: a small electric water heater can be installed either in the cabinet under the lavatory or under the floor of the house in the crawl space.

If the heater is in the crawl space, there is a risk of the heater's being frozen if it is turned off during very cold weather. The secret of preventing freezing is good maintenance. Close the vents in the crawl space during cold weather. Water heaters that operate on 110 volts are made in small sizes (5, 8, and ten gallons), and they can be switched off when no hot water is needed. To heat such a small quantity of water uses relatively little electricity, and the overall saving, both in initial cost and in operating cost since no water is wasted, makes this type of installation worthwhile. The electrical connection to the heater can be either a solid switched connection or a plug-in type. No matter which method is used, a substantial saving should be realized by the installation of a point-of-use heater.

Making the connection to the existing cold-water piping will depend on the type piping, galvanized pipe or copper tubing.

Instructions for soldering (sweating) copper tubing will be found in Chapter 31, Soldering Copper Tubing. After the connection has been made, the new piping can be extended to the new addition. Copper-plated pipe hooks, available at plumbing or building-supply stores, should be used to hang the copper tubing from the floor joists. Two ways to install point-of-use water heaters are shown in Figures 67 and 68.

Fig. 68   Water heater installed in crawl space.

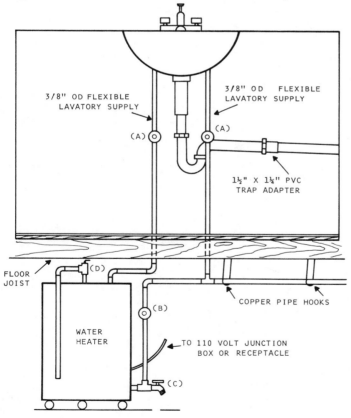

3/8" OD FLEXIBLE LAVATORY SUPPLY

3/8" OD FLEXIBLE LAVATORY SUPPLY

(A)

(A)

1½" X 1¼" PVC TRAP ADAPTER

FLOOR JOIST

(D)

COPPER PIPE HOOKS

(B)

WATER HEATER

TO 110 VOLT JUNCTION BOX OR RECEPTACLE

(C)

(A) 1/2" COPPER X 3/8" OD COMPRESSION STOP
(B) 1/2" COPPER SWEAT STOP
(C) 1/2" BOILER DRAIN VALVE
(D) 3/4" TEMPERATURE & PRESSURE RELIEF VALVE

Chapter Ten

Wiring the Addition

TYPES OF ELECTRICAL WIRING

It was common practice many years ago to install knob and tube wiring, as shown in Fig. 69, and in some parts of many older houses this type of wiring is still in use. With knob and tube wiring one wire is hot, the other neutral; there is no ground back to the fuse cabinet.

Fig. 69   Knob and tube wiring.

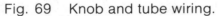

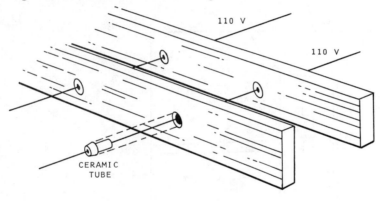

110 V

110 V

CERAMIC
TUBE

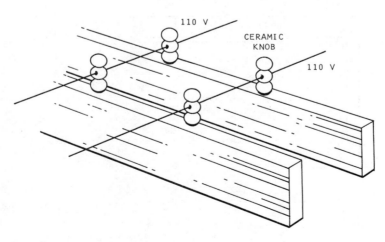

110 V

CERAMIC
KNOB

110 V

Some years later armored cable called BX came into general use, but BX has been outlawed in many areas because ground faults (shorts) often developed, causing the exterior cable to heat, in some cases resulting in fires.

Still later a nonmetallic cable was developed, carrying hot and neutral wires encased in a tough plastic sheathing. This type is available with or without a bare ground wire encased in the sheathing. Nonmetallic (NM or NMC) cable, often called Romex, is now the most widely used wiring for residential use. When connecting into existing systems containing a ground wire or on new work, you should always use cable containing a ground wire.

NM cable is used for exposed and concealed wiring in normally dry locations; NMC is a moisture- and corrosion-resistant cable which can be used in any location where NM cable is permitted and in addition can be used in damp, corrosive and moist locations. The National Electrical Code places certain restrictions on the use of both NM and NMC cable. They are not permitted in any multifamily dwelling that exceeds three floors above grade nor in anything larger than a two-family dwelling.

It is almost certain that any existing knob and tube or BX wiring is already overloaded. Neither of these types of wiring should be connected into or extended. Entire new circuits should be installed from the points of need to the fuse or breaker cabinet.

BX cable is shown in Fig. 70, NM cable is shown in Fig. 71.

Fig. 70   BX electrical cable.

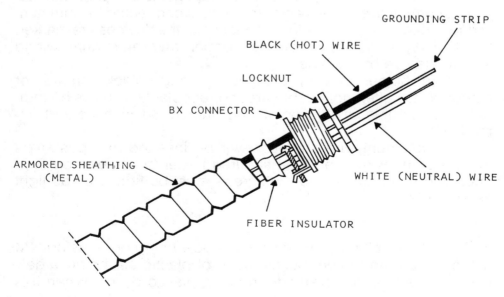

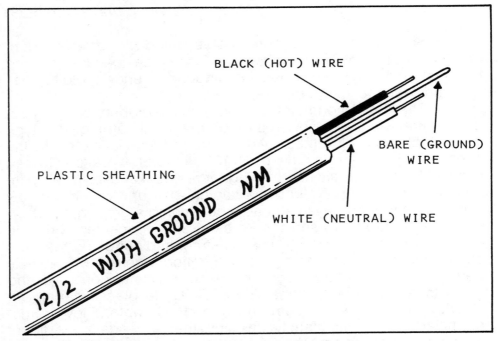

BLACK (HOT) WIRE

BARE (GROUND) WIRE

PLASTIC SHEATHING

WHITE (NEUTRAL) WIRE

12/2 WITH GROUND NM

Fig. 71   NM cable is the most used type for residences.

ALUMINUM WIRING

If your home has existing aluminum wiring and you add new wiring, the new wiring should be copper. Many problems have arisen through the use of aluminum wiring. It is a suspected cause of many home fires, and with a few exceptions it is now prohibited. If replacement of switches and receptacles connected to aluminum wiring is necessary, check to be certain that the devices are marked CO/ALR. Wire nuts connecting copper and aluminum wiring should also be of this type.

In homes having aluminum wiring, black streaks or smudges are often noticeable around receptacles. This is an indication of trouble at this point and should be investigated immediately.

Aluminum wire should never be inserted into push-type terminals; the wire should be secured under the screws on switches and receptacles and the screws must be tightened as tight as possible.

Almost all residential wiring is now being done using NM cable. Switch and outlet boxes made of plastic are being widely used. There is a disadvantage in using plastic boxes: When it is

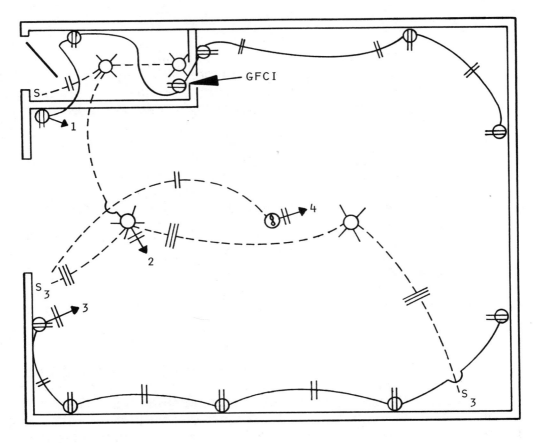

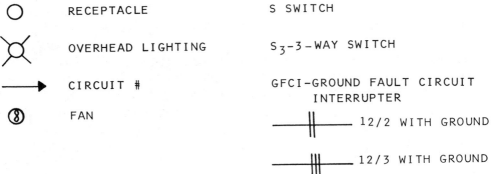

○	RECEPTACLE	S	SWITCH
⊗	OVERHEAD LIGHTING	S_3	3-WAY SWITCH
→	CIRCUIT #	GFCI	GROUND FAULT CIRCUIT INTERRUPTER
⊛	FAN	┤┤	12/2 WITH GROUND
		┤┤┤	12/3 WITH GROUND

Fig. 72 Electrical circuiting of the addition.

desirable to place two or three switches in line, if metal "gang"-type boxes are used, the right side of one box and the left side of another can be removed by loosening screws and the two or three boxes can be ganged together. If plastic boxes are used they cannot be joined together in this way and must be purchased as the right-size box for two, three or four switches.

Start your wiring project by marking the outlet and switch

box locations and nailing the boxes to studding and floor or ceiling joists. Table G shows the recommended heights for these boxes. Receptacle boxes should be spaced equal distances apart whenever possible. The outside edges of boxes should be flush

TABLE G—RECOMMENDED HEIGHTS FOR ELECTRICAL AND TELEPHONE OUTLETS

	ABOVE FINISHED FLOOR	ABOVE SINK AND LAVATORY COUNTER TOPS
SWITCHES	48" TO CENTER	8"
RECEPTACLES	14" TO CENTER	8"
THERMOSTATS	54" TO CENTER	
TELEPHONE OUTLETS (PREWIRED)	14" TO CENTER	6"

with the finished wall. The National Electrical Code, Section 210-24 to Section 210-25, requires that no point, measured horizontally along a floor line in any wall space, be more than 6 feet from an outlet in that space. This translates roughly that receptacles shall be no more than 12 feet apart. In kitchen and dining areas a receptacle outlet shall be installed in each counter space wider than 12 inches. If counter-top spaces are separated by refrigerators, range tops or sinks, each space shall be considered as a separate space and thus requires a receptacle outlet. In a bathroom at least one outlet must be installed adjacent to the basin location. This outlet should be protected by a ground fault circuit interrupter, as shown in Fig. 73. At least one outlet shall be installed in each basement and attached garage. A single-family dwelling shall have at least one outside receptacle. At least one receptacle shall be provided for the laundry (if laundry facilities are permitted and installed), and the receptacle shall be placed within 6 feet of the intended location of the appliance.

A small-appliance circuit may have six duplex receptacles. When wiring is being installed, it should be looped from one box to another, secured as shown in Fig. 74. A feed wire (electricians call it the "home run") is then extended from the nearest outlet box or junction box in the circuit to the service panel. At least 8 inches of extra wire should be left at each outlet for future connections.

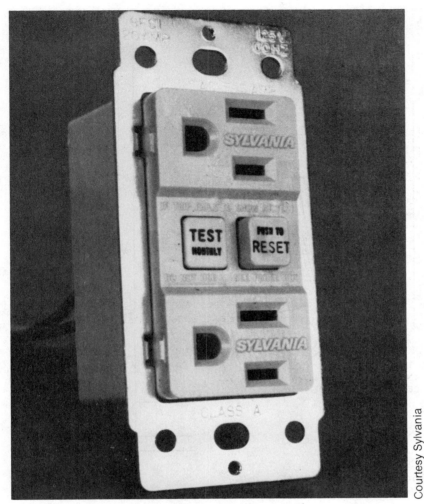

Fig. 73   A receptacle-type ground fault circuit interrupter.

When installing electrical wiring in walls or through ceiling joists, ¾-inch holes should be drilled as near the center line of the studs or joists as possible. Slack should always be left; the cable should never be pulled tight, especially at turns. In stud walls cables should be stapled at 54-inch intervals and within 8 inches of plastic boxes and 12 inches of metal boxes, as shown in Fig. 74. One common mistake in installing NM cable is not leaving enough wire protruding from the box to make future connections. While installing wiring, always leave at least 8 inches of wires extending from the boxes, as shown in Fig. 74. It's easy enough to cut off any excess wire, but it's very hard to make a connection if the wire is too short.

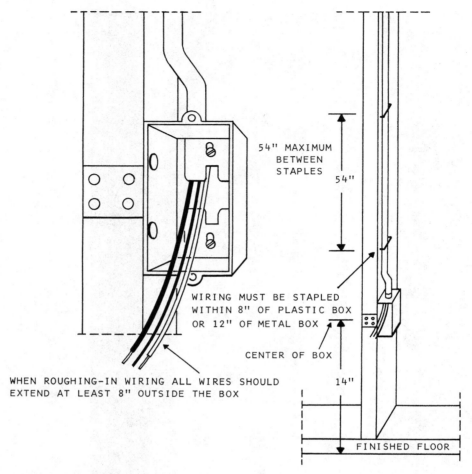

54" MAXIMUM
BETWEEN
STAPLES

54"

WIRING MUST BE STAPLED
WITHIN 8" OF PLASTIC BOX
OR 12" OF METAL BOX

CENTER OF BOX

WHEN ROUGHING-IN WIRING ALL WIRES SHOULD
EXTEND AT LEAST 8" OUTSIDE THE BOX

14"

FINISHED FLOOR

Fig. 74   Installing NM or NMC cable.

If you plan to install a ceiling fan, either now or at some future period, install the wiring and a 2 × 4 support, shown in Fig. 81, at the fan location. When you are roughing in the wiring for a ceiling fan, if the fan speed is controlled by a multispeed wall switch and if the fan is also to have an attached light fixture, two hot wires will be needed in the ceiling electrical box. The simplest way to install this wiring is to run two two-wire with ground cables from the wall switch location to the ceiling box. The reason two hot wires are needed is that if only one hot wire, controlled by a multi-speed wall switch is installed, and if the light fixture as well as the fan is connected to this wire, as the fan speed is increased, the light will be brighter, as the fan speed is decreased, the light will dim. Two single pole wall switches to be installed later will control the electric power to the fan location. One switch in series with the multispeed fan switch will control the fan, the other will control the power to the lighting fixture on the fan.

After the walls and ceiling have been closed in, it may be impossible to add the second hot wire, so it should be done when the wiring is roughed in. Even though it may not be used at this time, it will be available in the future. If it is not used at this time, the ends of each wire should be protected with wire nuts to prevent an accidental short within the box.

The electrical circuits shown in Fig. 72 indicate the locations of switches, receptacles, overhead lighting and a separate circuit, #4, for an attic fan. The lighting fixtures in the half bath are not on the same circuit as the receptacles; thus if power is lost on the receptacle circuit, lighting is still available. The ground fault circuit interrupter, GFCI, shown in Fig. 73, is installed in the receptacle circuit in the half bath, as required by the National Electrical Code. Ground fault circuit interrupters are "people protectors" designed to sense a very small current leakage and then trip a circuit breaker. GFCIs are made in two types: one serves as a receptacle; the other is a circuit breaker which mounts in the service panel and gives ground fault protection to all openings in the circuit. Wiring can seem very complicated when a breaker panel is opened, exposing a tangle of wires. If the original job was done properly the wires will run in a neat and orderly fashion. All black and red circuit wiring will connect to the *load* side of the circuit breakers, all white wires will connect to the neutral bar and all bare, green or ground wires will connect to the ground bar if the wiring was done correctly. Wiring connections in a service (breaker) panel are shown in Fig. 75. All black and red wires are hot, carrying current when the breakers are *on*.

The hot wires should always be connected to the dull or brass-colored screws on switches, receptacles or other devices. White wires connect to the bright or silver-colored screws. Bare (ground) wires connect to a green or hex-shaped screw. Single-pole switches should be wired with a black wire; a switch should always open (break) a hot wire in a circuit.

The correct way to connect a single-pole switch and receptacle is shown in Fig. 76. Note that only the black (hot) wire is connected to the switch; the white wires are wire-nutted together. Wiring to the receptacle uses both the black and the white wires, and the ground wire connects to the green or hex-head screw on the receptacle. Three-way switches which control the current to a light or receptacle from either of two different points are explained in the section Three-Way Switches.

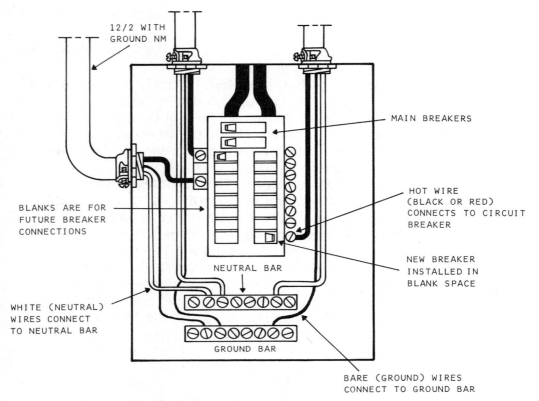

12/2 WITH
GROUND NM

MAIN BREAKERS

BLANKS ARE FOR
FUTURE BREAKER
CONNECTIONS

HOT WIRE
(BLACK OR RED)
CONNECTS TO CIRCUIT
BREAKER

NEW BREAKER
INSTALLED IN
BLANK SPACE

NEUTRAL BAR

WHITE (NEUTRAL)
WIRES CONNECT
TO NEUTRAL BAR

GROUND BAR

BARE (GROUND) WIRES
CONNECT TO GROUND BAR

Fig. 75   Wiring connections inside service panel.

Additional circuits can be added to existing fuse or
breaker panels if space is available and the incoming service is
large enough. The local electric utility company or a qualified
electrician can answer your questions in this area. If the electric
service is large enough but there is no room in the service panel, it
may be possible to add a subpanel containing circuit breakers for
additional circuit connections. Connections for a subpanel will
require the use of two circuits in the existing panel; therefore, the
capacity of the subpanel must be large enough to handle the new
circuits plus the two circuits to be transferred from the existing
panel. If the installation of a subpanel is necessary, I recommend
that it be installed by a qualified electrician.

After the rough-in wiring has been completed and before
the walls are covered, call for the rough-in inspection if one is
required.

Electrical wiring requirements will vary from one locality to
another. The building official or inspector in your area should be
consulted before you start work. All wiring must be installed according to local codes and ordinances.

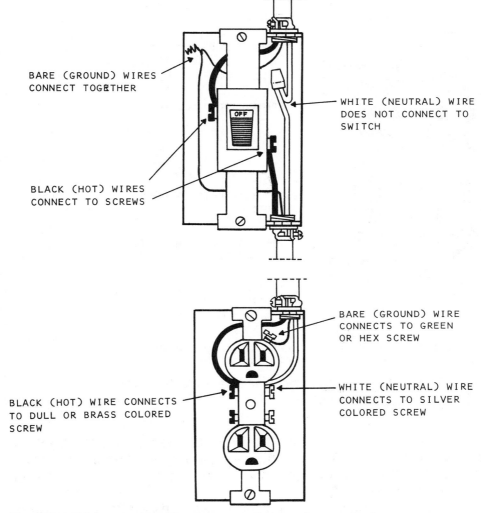

BARE (GROUND) WIRES
CONNECT TOGETHER

WHITE (NEUTRAL) WIRE
DOES NOT CONNECT TO
SWITCH

OFF

BLACK (HOT) WIRES
CONNECT TO SCREWS

BARE (GROUND) WIRE
CONNECTS TO GREEN
OR HEX SCREW

WHITE (NEUTRAL) WIRE
CONNECTS TO SILVER
COLORED SCREW

BLACK (HOT) WIRE CONNECTS
TO DULL OR BRASS COLORED
SCREW

Fig. 76   Wiring connections to a switch and receptacle.

ADDING A RECEPTACLE

It is often possible to extend an existing circuit and add one or two receptacles by connecting to an existing receptacle. First, turn the breaker controlling the existing circuit *off*, then double check to make certain that the circuit label is correct. Remove the cover plate, then remove the screws holding the receptacle and pull it out of the box. Connect the new wires, as shown in Fig. 77: black (hot) wire to the dull or brass-colored screw, white wire to the bright or silver screw and bare or ground wire to the green or hex screw or to the bare or ground wire already connected to the hex screw.

    After all connections are made, carefully push the receptacle back into the box and, using the screws removed earlier,

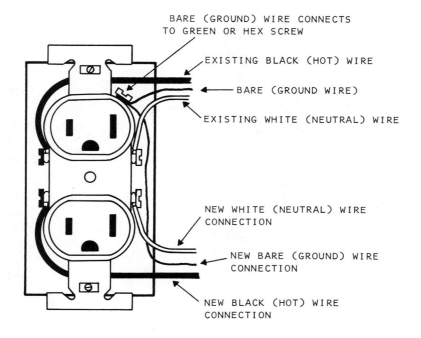

BARE (GROUND) WIRE CONNECTS
TO GREEN OR HEX SCREW

EXISTING BLACK (HOT) WIRE

BARE (GROUND WIRE)

EXISTING WHITE (NEUTRAL) WIRE

NEW WHITE (NEUTRAL) WIRE
CONNECTION

NEW BARE (GROUND) WIRE
CONNECTION

NEW BLACK (HOT) WIRE
CONNECTION

TO ADD ANOTHER RECEPTACLE CONNECT WIRING AS SHOWN
AND EXTEND TO NEW OUTLET CONNECTION

Fig. 77   Wiring connections to receptacle for extending
wiring to new outlet.

fasten the receptacle to the box. Replace the cover plate and turn the circuit breaker back on. The wiring at the new receptacle at the end of the circuit should be completed before you connect it to the existing receptacle.

Metal boxes used for NM cable have knockout plugs through which the cable enters the box and hold-down clamps to secure the cable. Remove the necessary knockout plugs before nailing the box to a wall stud. Strip off 8 inches of the plastic cable *sheathing (not the wire insulation)*, and insert the wires through the knockout holes into the box. Pull the wires into the box until about 1 inch of cable sheathing is under the hold-down clamps, then tighten the clamps just enough to hold the cable securely.

CIRCUIT BREAKERS

The safety, convenience and long-term economy of circuit breakers are good reasons for the increasing popularity of these superior devices. No longer do you have to first find the blown fuse, then search for a replacement only to find you don't have one. When a circuit breaker "trips," or opens a circuit, after the cause has been found and repaired it is only necessary to reset the breaker and the circuit involved is working again.

Blank spaces (knockouts) in breaker cabinets can be used to add one or more circuit breakers and circuits to the existing electrical systems.

Adding circuits requires working inside the service (breaker) cabinet; to do this the outer faceplate of the cabinet must be removed. *Before removing this plate turn the main breakers OFF.* The high-voltage connections and the ground are both exposed when the faceplate is removed and the combination can be deadly. Extreme caution is necessary when working inside the service cabinet: even though the main breakers are in *off* position, the electric power is still on up to the main breakers.

The Sylvania circuit breakers shown in Fig. 78 are the plug-on type. They are installed by hooking the *load* end (the end to which the circuit wire is to be attached) of the breaker under the hold down tab in the panel. Press the *line* end of the breaker down so that the plug-on clip slips over the bus stab. Insert breakers by pairs for easier installation.

Courtesy Sylvania

Fig. 78   Plug-on circuit breakers.

If you plan to add additional circuits to your existing breaker panel check to be sure that:

> 1. Space is available for one or more additional breakers;
> 2. You purchase a breaker that will fit the existing panel. If there is no main breaker in the breaker panel, do not attempt to do any work; call an electrician to check out the electric system.

When installing wiring from a service (breaker) panel for new circuits, it is recommended that you use wiring no smaller than number 12. Number 12 wire will handle 20 amps, whereas number 14 will only handle 15 amps. The cost of boxes, switches and receptacles is the same, and the cost differential between number 12 and number 14 wire is negligible.

FINAL WIRING CONNECTIONS TO A FIXTURE OR RECEPTACLE

Before starting on this work be certain that the electrical power to this circuit is *off*; do not trust a breaker label, it may be wrong.

Final electrical connections to a fixture or receptacle will require a single black (hot) wire and a single white (neutral) wire and may require a single bare or ground wire. These wires must be added to the existing black, white and bare wires already in the box. A typical wall or ceiling box may contain two black wires, two white wires and two bare (ground) wires. If a box contains a red wire it should be a "hot" wire. "Hot" wires are always connected together, neutral wires are always connected together and ground wires are always connected together. The connecting of hot (black or red) wires to white (neutral) or bare (ground) wires will create a "dead short" and blow the fuse or trip the breaker controlling the circuit.

The single connecting wires for the fixture or receptacle are prepared in this manner: Cut a black and a white wire, each six to eight inches long, and remove approximately three quarters of an inch of insulation from both ends of each wire. Remove the same amount of insulation from the black and white wires in the box. Add one end of the short black wire to the bared ends of the other black wires in the box and twist the bared ends together tightly, using pliers. Then twist a correctly sized wire nut onto the joint. Repeat this process with the white wires. Local electrical codes may require that the joints be soldered and taped with electrical tape instead of using wire nuts.

The bare (ground) wires should be twisted tightly together and if a grounded type receptacle or fixture is used, a short wire should be connected from the bare or ground wires to the fixture or receptacle as shown in Figures 77 and 79.

After the short connecting wires have been added to the grouped and connected black, white and neutral wires in the box, the fixture or receptacle can be installed.

If the new fixture or receptacle is being installed to replace an existing one the connections explained above will already have been made.

INSTALLING LIGHTING FIXTURES
When installing new fixtures or replacing existing fixtures, the very first thing to do is to make certain the electric power is off to the circuit you are working on. When a new fixture is unpacked, you will find a small envelope with the screws, nipples, locknuts and other assorted parts needed to install the fixture. There are several different ways to hang fixtures, depending on the type of fixture. Read the manufacturer's instructions and study the installation drawings furnished with the fixture.

Lightweight fixtures are usually installed as shown in Fig. 79. In some cases the mounting screws are screwed directly into

Fig. 79   Using a fixture strap to hang lightweight lighting fixture.

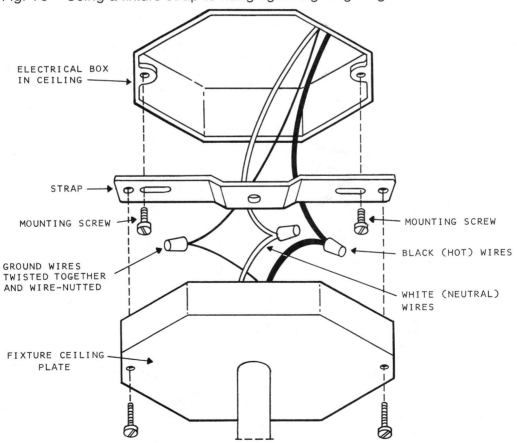

the ceiling electrical box; other types of fixtures may need a mounting strap. If a strap is used, the screws holding the strap will go into the box and the fixture will be attached to the strap.

Heavier fixtures are often installed as shown in Fig. 80. A fixture stud extends down through the top of the ceiling box. A fitting (called a hickey) is screwed to the stud and a nipple is attached to the hickey with locknuts. The skirt, or escutcheon, of the fixture is held against the ceiling by the collar screwed onto the nipple. The fixture stud is part of the hanger bar, indicated by the dotted lines, and is nailed to the ceiling joists when the electrical wiring is being roughed in.

Fig. 80   Typical method of hanging heavy lighting fixture.

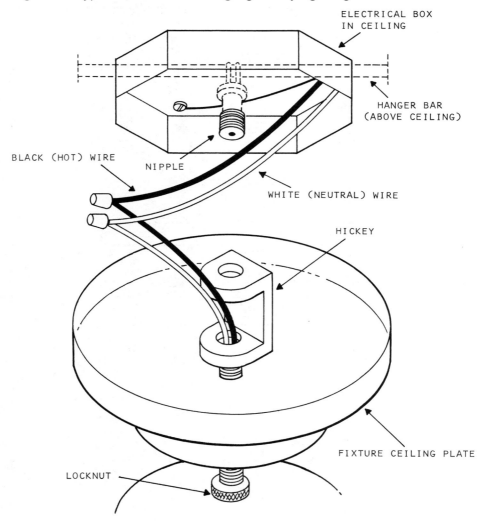

INSTALLING PADDLE FANS

Electrical wiring of a paddle fan is a simple matter of connecting the black and white wires from the fan to the black and white wires in the electrical box and connecting the ground wire from the box to the green or hex screw on the fan. Hanging the fan is often the problem. Most fans are made to hang from screw hooks, and the screws must be inserted into solid wood to hold the fan's weight. If the fan is to be installed on an existing ceiling, replacing an overhead light fixture (and if it's possible to get above the ceiling), a 2 × 4 can be nailed into the ceiling joists as shown in Fig. 81. Remove the center knockout in the electrical box and the screw hook can be screwed into the 2 × 4, as shown.

When it's not possible to get above the ceiling to mount the 2 × 4 blocks, an alternative fixture-hanging method is to use a stud finder, Fig. 106, to locate the ceiling joist. The screw hook can then be screwed into the bottom of the joist to hold the fan and the wiring can be run across the ceiling, using swag hooks, and

Fig. 81   Using a 2 × 4 support for a ceiling fan.

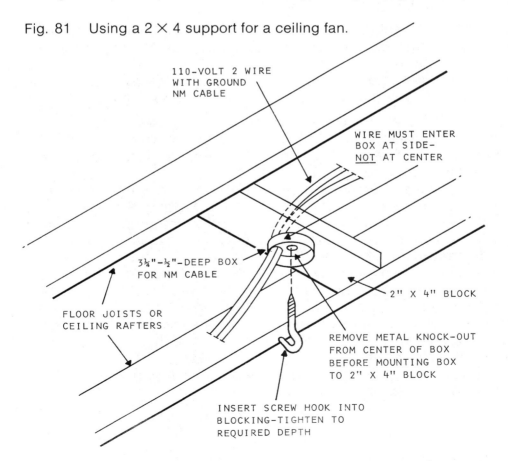

110-VOLT 2 WIRE
WITH GROUND
NM CABLE

WIRE MUST ENTER
BOX AT SIDE-
NOT AT CENTER

3¼"-½"-DEEP BOX
FOR NM CABLE

2" X 4" BLOCK

FLOOR JOISTS OR
CEILING RAFTERS

REMOVE METAL KNOCK-OUT
FROM CENTER OF BOX
BEFORE MOUNTING BOX
TO 2" X 4" BLOCK

INSERT SCREW HOOK INTO
BLOCKING-TIGHTEN TO
REQUIRED DEPTH

dropped down the nearest wall and plugged into a receptacle. Fans for this type of application are made with a speed-control switch mounted on the fan body and come equipped with installation instructions and hardware (screw hook, swag hook, etc.). When you are building an addition or a new house, the electrical box can be mounted on a 2 × 4 nailed between the ceiling joists, as shown in Fig. 81. Be sure to remove the center knockout in the box before the box is secured to the 2 × 4.

If the fan has a separate speed control, the control box should be mounted at the side of the switch that controls the overhead fixture box. Follow the manufacturer's instructions when installing the fan. Wiring must conform to local codes.

THREE-WAY SWITCHES

Two three-way switches are used to control one fixture or receptacle from two separate locations.

Examples:

1. A light in a stairway, controlled by one three-way switch at the top and also by another three-way switch at the bottom of the stairway.
2. A garage light, controlled by a three-way switch in the garage and also by another three-way switch in the house.

Three-way switches can be identified easily: they always have three wires connected to them, *none* of which is a ground (bare or green wire). There are three types of switches: mercury type (the best), quiet type and snap type.

To replace a three-way switch, first remove the fuse or turn off the circuit breaker controlling the circuit. Remove the wall plate and switch. Remove the wire connected to the common terminal of the old switch. (This screw may be marked "common" or may be a different color from the other two screws.) Connect this wire to the common terminal or copper-colored screw on the new switch. Remove the other two wires from the old switch and connect them to the remaining two screws on the new one. If the new switch has push-in terminals, these may be used if the existing wiring is copper. Mount the new switch in the box, replace the cover plate and turn the power back on. If the switch does not work properly the common wire is probably connected to the wrong terminal. Turn the power off again (fuse or circuit breaker), and check the wiring.

A wiring diagram of three-way switches is shown in Fig. 82. Circuits are usually marked on electrical plans with arrows and numbers. The arrow shows the feed wire to the circuit (the home run), followed by the circuit number. As shown on our floor plan, Fig. 72, circuit number 2 feeds overhead lighting in the main room and the half bath. The feed wire to the branch circuit, number 2, is number 12/2 with-ground NM cable. The wiring between the two three-way switches is number 12/3 with ground cable which contains one black, one red and one white wire and a bare ground wire. When

Fig. 82   Schematic and pictorial wiring diagrams showing how three-way switching works.

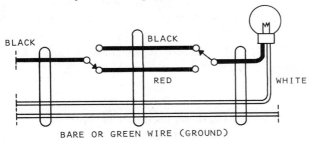

SCHEMATIC OF 3 WAY SWITCH INSTALLATION. AS SHOWN, LIGHT IS OFF. MOVE EITHER OF THE ARROWS, TURNING SWITCH ON AND LIGHT WILL BE ENERGIZED.

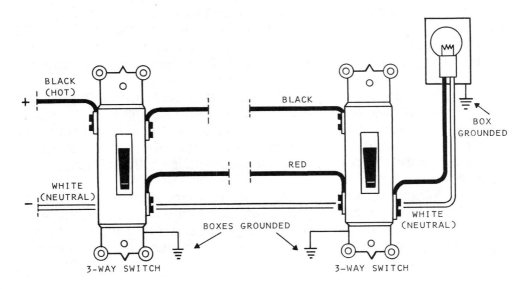

PICTORIAL WIRING OF 3-WAY SWITCHES

replacing three-way switches you may find a red, a white and a black wire connected to the switch instead of two blacks and a red, as shown in Fig. 82. White wires are identified as neutral wires and if, as in this case, a white wire is used as a hot wire, the white insulation should be painted black or wrapped with black tape, close to the connection point, to identify it as a hot wire.

Note: If your house is wired with aluminum wiring, when replacing switches or receptacles *use only devices marked CO/ALR*.
Push-in terminals are used on many switches and receptacles. These devices have a strip gauge on the back side to show how much insulation to strip from the wire, leaving the bare end. After stripping the wires to the length shown, insert the bare end into the terminal. To release the wires from this type terminal, insert a small screwdriver into the release slot and remove the wire.
Do not use aluminum wire in a push-in terminal.

THE ELECTRICAL SYSTEM—TROUBLESHOOTING

Continuity testers are available in electrical departments of building-supply stores and in hardware stores. One type uses a light bulb which will light when a circuit is completed. Another type uses a bell or buzzer which sounds if the circuit is complete. A complete and unbroken circuit means that a fuse or switch is good or that a wire between two points is unbroken: either will permit electrical current to flow in the circuit. When a fuse "blows," the link, or wire connection, through the fuse melts, or "opens," and current cannot flow through an open circuit.
In Fig. 83 the continuity tester is used to check fuses and switches. Whenever a continuity tester is to be used for this purpose, the fuses or switches *must* be removed or disconnected from the electrical circuits. When these tests are made, if the light does not light, the component being tested is faulty and should be replaced.
When replacing defective receptacles always use the correct type. Two types of receptacles are shown in Fig. 84. One type (A) is not a grounding type. The A type must *not* be replaced by the grounding type *unless* the box in which the receptacle is mounted

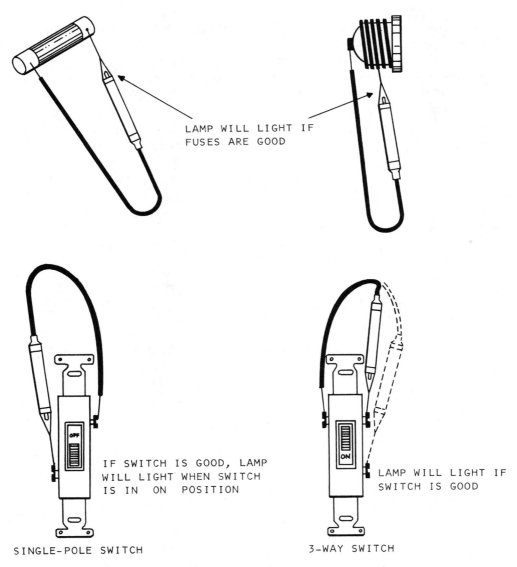

LAMP WILL LIGHT IF
FUSES ARE GOOD

IF SWITCH IS GOOD, LAMP
WILL LIGHT WHEN SWITCH
IS IN ON POSITION

LAMP WILL LIGHT IF
SWITCH IS GOOD

SINGLE-POLE SWITCH

3-WAY SWITCH

Fig. 83   Continuity tester can be used for troubleshooting.

is grounded. Three tests are shown in Fig. 85. One shows how to test for a ground on grounded-type receptacles. A neon voltage tester, available at hardware stores for about $1.00, and similar in appearance to the tester shown in Fig. 85, will glow when inserted into the hot and neutral slots of the receptacle if the receptacle is energized. If you are in doubt as to which is the hot slot, test from each slot to the ground screw; if the box is properly grounded, the tester will glow when one prong of the tester is in the hot slot and the other is grounded. If the tester does not glow when the center or ground screw is in contact, the box is not grounded.

Fig. 84 The difference between grounded and ungrounded receptacles.

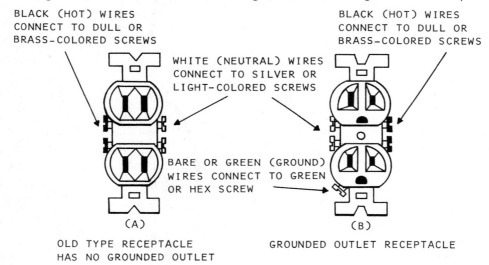

BLACK (HOT) WIRES
CONNECT TO DULL OR
BRASS-COLORED SCREWS

BLACK (HOT) WIRES
CONNECT TO DULL OR
BRASS-COLORED SCREWS

WHITE (NEUTRAL) WIRES
CONNECT TO SILVER OR
LIGHT-COLORED SCREWS

BARE OR GREEN (GROUND)
WIRES CONNECT TO GREEN
OR HEX SCREW

(A)

(B)

OLD TYPE RECEPTACLE
HAS NO GROUNDED OUTLET

GROUNDED OUTLET RECEPTACLE

A TWO-SLOTTED RECEPTACLE MUST NOT BE REPLACED WITH A GROUNDED RECEPTACLE
UNLESS THE BOX IS GROUNDED

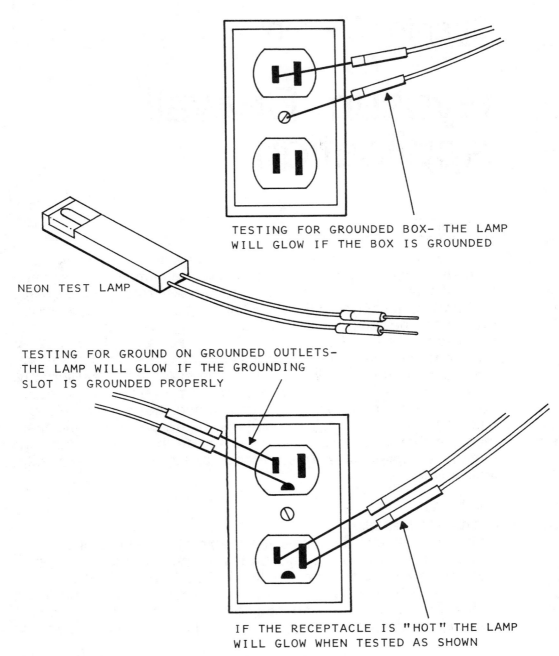

TESTING FOR GROUNDED BOX- THE LAMP
WILL GLOW IF THE BOX IS GROUNDED

NEON TEST LAMP

TESTING FOR GROUND ON GROUNDED OUTLETS-
THE LAMP WILL GLOW IF THE GROUNDING
SLOT IS GROUNDED PROPERLY

IF THE RECEPTACLE IS "HOT" THE LAMP
WILL GLOW WHEN TESTED AS SHOWN

Fig. 85   How to use a neon test lamp.

Chapter Eleven

Gypsum Drywall Application

PLANNING THE JOB

A little thought and planning before you start your project can result in a better-appearing job and a saving in material and time. Make a sketch of the areas to be surfaced with gypsum wallboard and lay out the board panels. Install the boards across (perpendicular to) the joists or studs. Use as long a board as can be handled to eliminate or reduce end joints. For example, in a 12 × 13-foot room where the ceiling joists run parallel to the 13-foot dimension, it is desirable to have the boards be 12 feet long. If they are 8 feet long, an end joint would be necessary in each course. Where end joints cannot be avoided, they should be staggered. It is usually better to apply the board on the ceiling first, then the sidewalls. It is often easier to use the adhesive/nail-on method of application. This method results in a higher-quality installation.

Estimating Materials

Using your sketch, determine the length and number of boards required. The nails, joint compound and tape needed can be estimated using the chart on estimating materials. After the wallboard is installed, the flat joints and the inside corners are to be reinforced with a paper tape and joint compound. The outside corners are to be reinforced with a gypsum-board metal corner bead and joint compound.

In the adhesive/nail-on method, Georgia-Pacific gypsum board adhesive is applied to the joists and studs before each piece of wallboard is positioned and nailed. The adhesive is applied to the framing member from a caulking gun in a bead about

⅜ inch in diameter. For each 1,000 square feet of wallboard use eight quart-size tubes of adhesive.

Tools Required
The basic tools you will need are

1. Wallboard cutting knife and heavy-duty knife blade
2. Wallboard hammer or carpenter's claw hammer
3. 4-foot T square or steel straightedge
4. Steel tape measure
5. Utility saw or keyhole saw
6. Utility finishing knives—4-inch and 10-inch blades
7. Plastic pan for joint compound
8. Sandpaper (medium grade) for joint finishing
9. Caulking gun

CUTTING GYPSUM WALLBOARD

Use the T square or straightedge and wallboard knife for scoring, as shown in Fig. 86. With the knife at right angles to the board, score completely through the face paper. Then apply firm, even pressure to snap the board. Fold back the partially separated portion of the board and use the knife again to cut the back paper. Rough edges should be smoothed. Panels can be cut with a saw if desired.

Georgia-Pacific Ready-Mix is a premixed, ready-to-use product that can be used for the complete job of taping, filling, spotting, nailheads and finishing. The chart shown below indicates approximately how much joint compound and tape will be needed for your job.

Estimating Materials

Wallboard Thickness	Nail Type	Approx. lbs. per sq. of gypsum wallboard
½″	1⅝″ coated gypsum-board nail	5¼ lbs.
⅝″	1⅞″ coated type gypsum-board nail	5¼ lbs.

Fig. 86   T square serves to guide knife when cutting drywall.

The following chart will tell you approximately how much joint compound and tape will be needed.

Estimating Ready-Mix Joint Compound and Tape

Georgia-Pacific Gypsum Wallboard sq. feet	Estimated Amount of Georgia-Pacific Ready-Mix Joint Compound	Estimated Amount of Georgia-Pacific Wallboard Tape
100–200 sq. ft.	1 gal.	Two -60 ft. rolls
300–400 sq. ft.	2 gals.	Three -60 ft. rolls
500–600 sq. ft.	3 gals.	One -250 ft. roll
700–800 sq. ft.	4 gals.	One -250 ft. roll One -60 ft. roll
900– 1000 sq. ft.	One 5-gal. pail	One -250 ft. roll Two -60 ft. rolls or One -500 ft. roll

Cutting Openings

It will be necessary to cut holes in the wallboard for electrical outlets, light receptacles, etc. The distance of the opening from the end and edge of the board should be carefully measured and marked on the face of the wallboard. The opening should then be outlined in pencil and cut out with a keyhole saw. The cutout must be accurate or the cover plate will not conceal the hole.

Nailing

Ceilings: With joists 16 inches on center, using ½-inch gypsum board, the nails should be at 7-inch intervals. With joists on 24-inch centers, using ⅝-inch gypsum board, the nails should be at 7-inch intervals, as shown in Fig. 87.

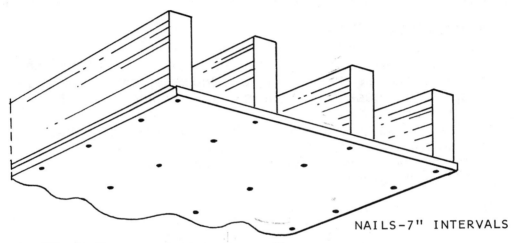

NAILS—7" INTERVALS

Fig. 87   Nailing gypsum drywall on ceiling.

Walls: For studs 16 inches on center, boards should be nailed at 8-inch intervals, as shown in Fig. 88.

Dimpling: "Dimple" all nails (Fig. 88) with a hammer blow firm enough to indent the board's face paper. (Don't break the paper.)

CEILING INSTALLATION
It is more difficult to install the ceiling boards because of the overhead positioning. It is desirable to have T braces to hold the board in place while it is being nailed. A satisfactory T brace consists of a 2-foot piece of 1 × 4 nailed onto the end of a 2 × 4. The length should be about 1 inch longer than the floor-to-ceiling height. When the adhesive/nail-on method is used, the edges should be nailed. All edges should be supported on framing. The nails should be driven to bring the board tight to the framing, then another blow struck to dimple the nail; be careful not to break the face paper.

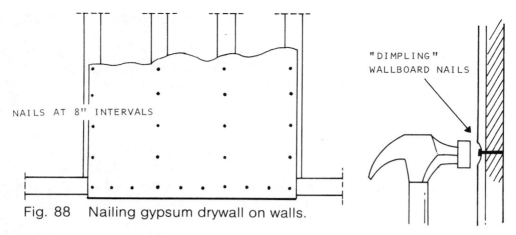

NAILS AT 8" INTERVALS

"DIMPLING"
WALLBOARD NAILS

Fig. 88   Nailing gypsum drywall on walls.

WALL APPLICATION

For horizontal application on sidewalls, install the top board first. Push the board up firmly against the ceiling and nail, placing nails 8 inches apart, as shown in Fig. 88. One exception, however, is to keep all nails back 7 inches from interior ceiling angles. Nails in the interior angles are quite apt to pop. If the adhesive/nail-on method is used, all of the field nailing can be eliminated. The nailing is around the edges of the board. If the board is bowed out in the center, it may be advisable to secure with a temporary nail until the adhesive sets.

Fig. 89   How to use T braces.

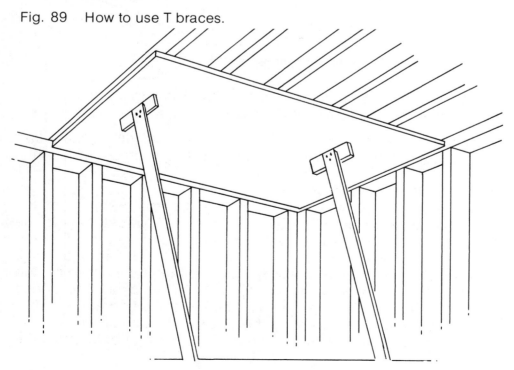

A vertical application places the long edges of the wallboard parallel to the framing members. This is more desirable if the ceiling height of your wall is greater than 8 feet 2 inches or the wall is 4 feet wide or less. Nailing recommendations are the same as for horizontal application.

Fig. 90 Applying gypsum drywall to stud wall.

Courtesy Georgia-Pacific Corp.

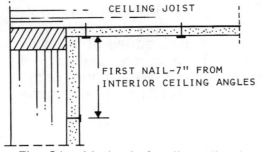
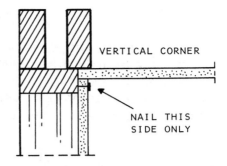

CEILING JOIST

FIRST NAIL—7" FROM
INTERIOR CEILING ANGLES

VERTICAL CORNER

NAIL THIS
SIDE ONLY

Fig. 91   Method of wall application.

Metal Corner Bead

To protect outside corners from edge damage, install metal corner bead after you have installed the wallboard. Nail the metal corner bead every 5 inches through the gypsum board into the wood framing, as shown in Fig. 92.

Joint Finishing

A premixed material such as Georgia-Pacific Ready-Mix Joint Compound is the easiest to use to finish joints, corners and nail-heads. A minimum of three coats of Ready-Mix is recommended for all taped joints. This includes an embedding coat to bond the tape and two finishing coats over the tape. Each coat should dry thoroughly, usually for 24 hours, so that the surface can be easily sanded. When sanding, wrap your sandpaper around a wood sanding block so you sand the surface evenly. Do not oversand or sand the paper surface. This may outline the joint or nailhead through the paint.

Fig. 92   Installing metal corner bead.

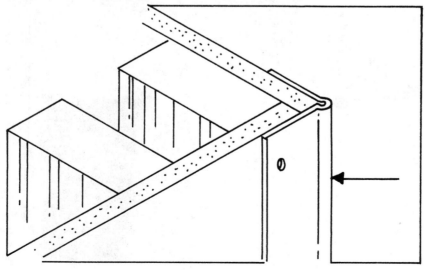

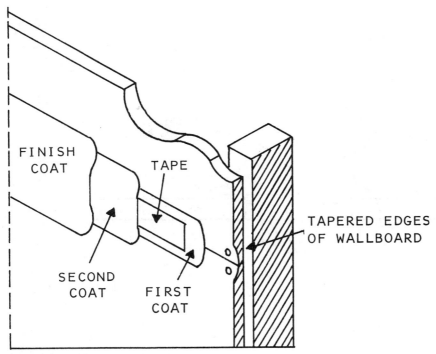

FINISH COAT

TAPE

TAPERED EDGES OF WALLBOARD

SECOND COAT

FIRST COAT

Fig. 93   Steps in joint finishing.

Applying Bedding Compound
Take your 4-inch joint-finishing knife and apply the Ready-Mix Joint compound fully and evenly into the slight recess created by the adjoining tapered edges of the board.

Applying Wallboard Tape
Next, take your wallboard tape, center it over the joint and press the tape firmly into the bedding compound with your wallboard knife held at a 45-degree angle. The pressure should squeeze some compound from under the tape, but enough must be left for a good bond, as shown in Fig. 95.

Applying Finishing Coats
When thoroughly dry—at least 24 hours—apply a fill coat extending a few inches beyond the edge of the tape and feather the edges of the compound. When the first finishing coat, shown in Fig. 96, is thoroughly dry, use your 10-inch joint-finishing knife and apply a second coat and feather the edges about 1½ inches beyond the first coat. When this coat is dry, sand lightly to a smooth even surface. Wipe off the dust in preparation for the final decoration. Total width should be from 12 to 14 inches. Sanding of the joints is shown in Fig. 96.

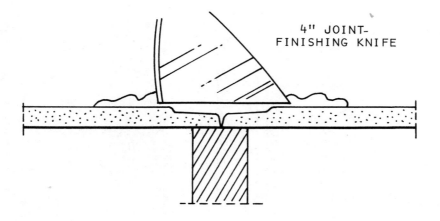

4" JOINT-
FINISHING KNIFE

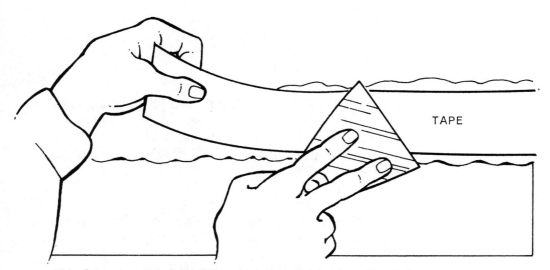

TAPE

Fig. 94   Applying bedding compound and wallboard tape.

Fig. 95   The correct way to apply wallboard tape.

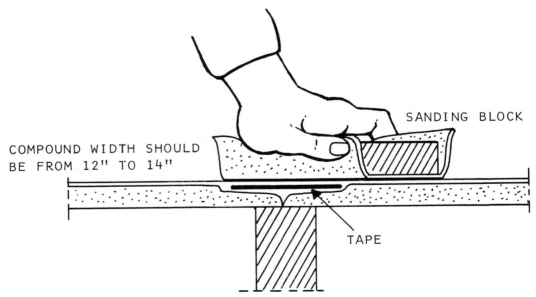

SANDING BLOCK

COMPOUND WIDTH SHOULD
BE FROM 12" TO 14"

TAPE

Fig. 96   Applying and sanding the finishing coat.

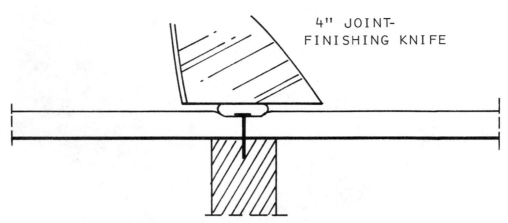

4" JOINT-
FINISHING KNIFE

Fig. 97   Finishing nail holes.

Fig. 98   Finished nail hole is flush with wallboard surface.

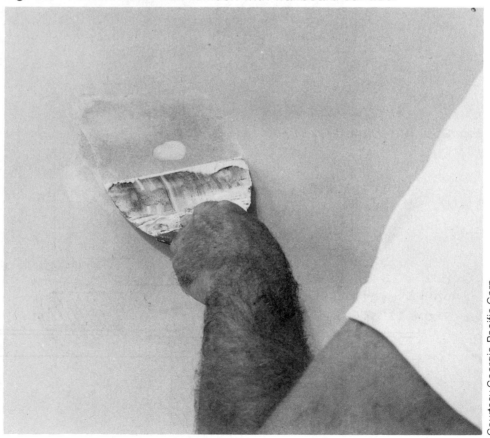

Finishing Nailheads

Draw your 4-inch joint-finishing knife across nails to be sure they are below the surface of the board (Fig. 97). Apply the first coat of Ready-Mix with even pressure to smooth the compound level with the surface of the board. Do not bow knife blade with excess pressure, as this tends to scoop compound from the dimpled area. When first coat is dry, apply second coat, let dry, sand lightly and apply third coat. Sand lightly before applying your decoration. An additional coat may be needed, depending on temperature and humidity.

Finishing End Joints

You use basically the same steps with end, or butt, joints as you do with tapered edges. (The two long edges are tapered slightly to aid in joint finishing. The two short or cut edges require more feathering with a larger putty knife or trowel.) The end joints are not tapered, so care must be taken not to build up the compound in the center of the joint. This encourages ridging and shadowed areas. Feather the compound well out on each side of the joint. The final application of joint compound should be 14 to 18 inches wide, as shown in Fig. 99.

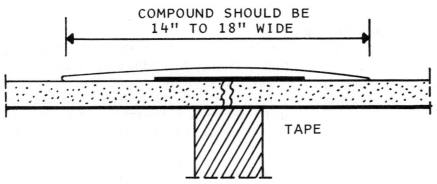

Fig. 99   How to finish end joints.

Finishing Metal Corner Bead

Be sure the metal corner is attached firmly. Take your 4-inch finishing knife and spread the Georgia-Pacific Ready-Mix 3 to 4 inches wide from the nose of the bead, covering the metal edges. When it is completely dry, sand lightly and apply the second coat, feathering edges 2 to 3 inches beyond the first coat. A third coat may be needed depending on your coverage. Feather the edges of each coat 2 or 3 inches beyond the preceding coat. When applying the compound hold the finishing knife against the nose of the bead.

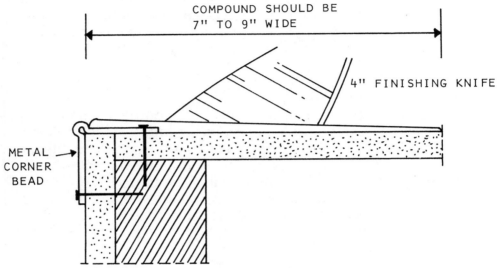

Fig. 100   Finishing metal corner bead.

Finishing Inside Corners

Cut your tape the length of the corner angle you are going to finish. Apply the Georgia-Pacific Ready-Mix with your 4-inch knife evenly about 1½ inches on each side of the angle. Use sufficient compound, as shown in Fig. 101, to embed the tape. Fold the tape along the center crease and firmly press it into the corner. Use enough pressure to squeeze some compound under the edges.

Fig. 101   Finishing inside corners.

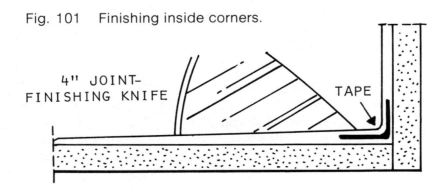

Feather the compound 2 inches from the edge of the tape. When first coat is dry, apply a second coat. Feather the edge of the compound 1½ inches beyond the first coat. Apply a third coat if necessary, let dry and sand to a smooth surface. Use as little compound as possible at the apex of the angle to prevent hairline cracking.

Final Decoration

After the joint treatment is thoroughly dry, all surfaces must be sealed or primed with a vinyl or oil base primer/sealer. This equalizes the absorption difference between the exposed surface and the joint compound surface. You will then have a uniform texture and suction over the entire wall or ceiling. When the primer/sealer has dried, apply your final decoration according to the manufacturer's instructions.

A water-resistant wallboard should be used in bath or shower areas, where water penetration through the wall will damage it. Georgia-Pacific Tile Backer Board is a water-resistant wallboard which serves as a backer for ceramic tile in bath or shower areas.

HARDWOOD FLOORING

At this point our addition has only the subflooring applied. If hardwood flooring is desired it is available in random lengths, either unfinished or prefinished. Prefinished hardwood parquet flooring is available in several different size squares, depending on the manufacturer. If unfinished flooring is used, finishing the floor requires the following steps:

1. laying the flooring and setting the nails
2. applying wood filler and allowing it to dry
3. sanding flooring surface
4. removing dust and sealing flooring with shellac, varnish or other desired type sealer

Prefinished parquet flooring is applied using a special mastic made specifically for this purpose. It will require only an occasional cleaning and waxing if properly cared for.

Random length prefinished flooring is applied with nails to a wood subfloor, the nails are concealed in the joints. The supplier can furnish complete installation instructions when the flooring is purchased.

FLOOR COVERINGS

Carpet, vinyl tile or roll type vinyl coverings are suitable for application over solid subflooring. Installing carpet requires in most cases, the use of specialized tools and material not ordinarily available to most do-it-yourselfers. Because labor is only a small part of the cost of installing carpet I recommend that when carpet is purchased installation be included.

The installation of vinyl type floor coverings is explained in another section of this book.

Chapter Twelve

Painting

INSIDE PAINTING

Gypsum Drywall and Plaster

Latex paints are favored for all areas of the home except bathrooms and kitchens. Because of moisture and grease, the preferred paint for bathrooms and kitchens is high-gloss or semigloss enamel. Also, semigloss and high-gloss enamels reflect more light and seem brighter and more cheerful in these areas. White, as an example, will reflect 80 percent of the available light while a light blue or light green will reflect only 40 percent of the available light.

Half of the work of painting is getting ready. Moving the furniture out of the room is best; what can't be moved out should be covered with drop cloths. Use a drop cloth to cover the floor too. The time spent in preparation will be repaid in time not spent in cleaning up after the job is done. Cracks or holes in the walls or ceilings should be repaired with a drywall joint compound before starting the painting. Remove the plates covering electrical switches and receptacles, brass ceiling plates of electrical fixtures and anything else attached to a wall or ceiling which is removable. After painting, don't be in a hurry to replace these items: the paint will not dry completely for at least 12 hours.

The ceiling should be painted first, and if the walls are to be the same color, paint the top 2 inches of the wall when you come to it. A roller with a heavy nap should be used, and I find that a 2-inch-wide brush works best for corners, door and window trim and baseboards. New walls should be primed with a latex primer and sealer. Patched areas may require more than one coat to cover.

When repainting old walls, if the new paint is a light color going on over a dark color, apply a flat white primer; if the dark color still shows through, a second coat of flat white primer will be needed.

Painting a Wallpapered Room

Painting over wallpaper is not desirable. The moisture from the paint will often loosen the paper, causing unsightly bulges. The paper should be removed either by use of a steamer or by wetting the wall and using a scraper to remove the old paper. Steamers can be rented at hardware stores and tool-rental warehouses. Follow the instructions for starting the steamer; when steam has been generated, hold the pan against the wall at one spot for 30 or 40 seconds. After this time it should be possible to peel the paper from the wall. A scraper, preferably a 4- or 5-inch-wide flexible-blade type, can be used to remove any remaining paper or wallpaper paste. After the wallpaper has been removed, nicks, gouges or cracks will probably have to be repaired before painting; drywall joint compound is excellent for this purpose. Before starting to paint, go over the whole area with a medium grade of sandpaper to remove any bits of paper still remaining, then wipe down the wall with a damp cloth to remove any dust.

In kitchens and bathrooms, before applying a gloss enamel, the walls should be washed with a trisodium phosphate solution (Spic 'n Span, Oakite or Soilax) to remove any grease or soap film. If there is a gloss still on the old paint, the surface should be sanded lightly to remove the gloss. Scratches or dents in the wall can be repaired with drywall joint compound. Any patches or repairs should be thoroughly dry before painting.

Concrete-block or Poured-concrete Walls

Concrete-block or poured-concrete walls are made with portland cement, which contains alkali. An alkali-resistant paint should be used; your paint supplier can furnish latex paints made for use on concrete walls. If the walls have been painted before, they should be washed with a trisodium phosphate solution and any loose particles removed with a stiff wire brush.

OUTSIDE PAINTING

Gutters and Downspouts

Aluminum gutters and downspouts have been used for many years and normally require no painting since they do not rust. They can be painted to match the outside trim of the house. Galvanized gutters can be preserved by painting the inside with asphalt paint, and the outside can be painted to match the trim. Painting downspouts in place can be tedious. Care must be taken not to get paint on the house walls. It is generally easier to take a little extra time and care and paint the downspouts in place rather than remove them. Galvanized-metal roof flanges should be painted to keep them from rusting. An asphalt paint or rustproofing paint, such as Rustoleum, should be used.

Outside House Surfaces

Preparation for the job is more important than application of the paint. Inspect door and window casings. If cracks have developed between the casings and the buildings, use a caulking gun and glazing compound to fill the cracks.

If putty is loose or missing around windows, it should be replaced with new putty or glazing compound. Old paint which is chipped or cracked should be sanded smooth or removed. When removal is necessary, a propane torch using a flame-spreading burner is very useful. Direct the flame against the paint; when it has softened, a sharp putty knife can be used to remove it. Paint scrapers in different widths are helpful in removing old paint; also, wire brush attachments used in an electric drill will do a good job. Drive any loose nails in tightly and fill any knotholes or cracks with plastic wood.

If surfaces are moldy or mildewed, scrub the area with a trisodium phosphate solution and allow to dry thoroughly before painting. Mildew-prevention chemicals are available at paint stores and should be added to paint when mildew growths are a problem.

When paint has peeled off, it indicates that moisture has entered the surface from the unpainted side. The source of the moisture must be found and corrected to prevent the peeling or flaking from occurring again.

Before starting outside painting, cover shrubs, grass areas, porches and walks with a drop cloth to eliminate time-consuming cleanup.

Long-napped paint rollers can be used with many types of paint to give a smooth, professional finished appearance.

Finally, if you use a roller, you may want to try something I learned years ago from a professional painter. Tear off a piece of aluminum foil large enough to line the paint tray, with an inch or two extra to turn over the edges of the tray. Press down tightly into the tray and at the corners before turning the edges down. When you're through painting, lift out the foil. Tray-cleaning time is eliminated.

Follow the paint manufacturer's application instructions and mix the paint thoroughly. When oil-based paints are used, the surface must be clean and dry. When more than one coat is needed, the previously applied coat must be thoroughly dry before the next coat is applied. New paint applied to paint which has not dried will cause the paint to crack.

Chapter Thirteen

Heating and Cooling the Addition

Several factors must be considered when you are deciding how to heat and air-condition an addition to an existing home. Does the present heating and air-conditioning system have the capacity to handle the additional load? Assuming that the present system has the capacity, how would the costs of extending the present system balance against the cost of an independent system for the addition?

In all likelihood, the present heating and air-conditioning system will not handle the additional load. Builders are not noted for oversizing original equipment.

Assuming that the existing system could handle the additional load and depending on the location of the equipment and working conditions of the job, it is very likely that the cost of extending the present system will be much higher than the initial cost of an independent system.

Our 20 × 24-foot addition is not closed off from the original building and will only require supplemental heating and air conditioning. Electric baseboard radiation, one unit under each window on the 24-foot walls, should provide adequate supplemental heat. The length of the units and the wattage required will depend on the severity of the cold weather in your area and the types and amounts of insulation used in construction of the addition.

The initial cost of electric baseboard radiation is very reasonable, but to this cost must be added the expense of installing electrical wiring and receptacles for the 240-volt 60-hertz units and the cost of thermostats and installation.

To get an idea of the operating costs, let's assume that you're in an area where electricity, including fuel-adjustment charges, costs $.075 per kilowatt-hour. At this rate a 2-kilowatt (a kilowatt is equal to 1,000 watts) element would cost $.15 per hour to operate.

A small unit of electric baseboard radiation will be adequate for the half-bath area.

A 20 × 24-foot room with 8-foot-high ceiling contains 3,840 cubic feet of air. A 2,000 cf/m through-the-wall type air-conditioning unit will change the air in the addition every two minutes. The operating efficiency of the unit will depend on the type and quantity of the insulation used in the outside walls, ceiling and floor.

An energy-saving fireplace is an excellent source of supplemental heat. The fireplace shown in Fig. 175 is an energy-saving type which heats and circulates the room air. The installed cost of this fireplace would probably be less in most cases than the cost of extending the present heating system into the new room, and the fireplace will add value, far exceeding its cost, to the home.

Chapter Fourteen

Wall Paneling

Installing wall paneling is one of the easiest do-it-yourself projects the homeowner can tackle. Paneling can be applied either to wainscot height (36 inches) or to full ceiling height. Paneling adds warmth and richness to a room, and should, with a minimum of care, last the lifetime of the home. Both the cost and the inconvenience of painting and/or wallpapering are eliminated when paneling is applied. If your walls are 8 feet high and in good condition paneling is simply a matter of following standard application procedures. New walls, rooms with unusual architectural features or uneven walls, as in basements with concrete block or poured concrete walls, present no problem if you know how to go about the job.

PANELING EXISTING WALLS

Start by measuring the room and making a plan or sketch; it need not be complicated but it should be accurate and complete. A simple sketch of a room is shown in Fig. 102. Make the sketch on graph paper and add up the total footage; the chart in Fig. 103 shows the number of panels needed without deductions, then deduct for all openings. The resulting figure will be the number of panels needed. As shown in Fig. 102, after measuring perimeter of room, make deductions for doors, windows and fireplaces as follows:

> For a door (A) deduct ½ panel.
> For a window (B) deduct ¼ panel.
> For a fireplace (C) deduct ½ panel.
> For double (French) doors or double windows, count each door or window as a separate item.

If the perimeter of the room falls between the figures in Fig. 103, use the next higher number to determine panels required. These figures are for rooms 8 feet or less in height. For higher walls, add in the additional materials needed above the 8-foot level.

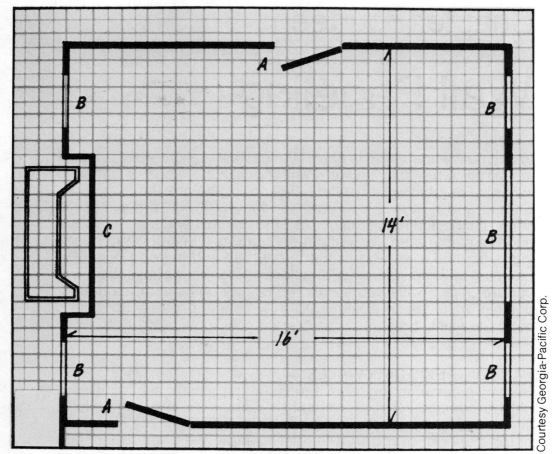

Fig. 102   How to figure deductions for openings in rooms.

Paneling does not require a lot of expensive tools; you probably have most of the tools you will need. For accurate measurements, a retractable tape measure 8 feet or longer and a large square (carpenter's framing square, 16 × 24 inches) and a level will do the job. For trimming the paneling, you will need a power saw with a sharp fine-toothed blade, a drill and a saber saw to make cutouts for electric receptacles, switches, etc. For the installation work you will need a hammer and nail set. If you are gluing the panels in place, you will also need a caulking gun and padded wooden block. The gluing-and-nailing method is recommended for a really good job. An inexpensive miter box to assist in making accurate molding joints and a pair of sawhorses to support the paneling while measuring and cutting complete the list of tools and materials needed.

Perimeter	No. of 4' x 8' panels needed (without deductions)
36'	9
40'	10
44'	11
48'	12
52'	13
56'	14
60'	15
64'	16
68'	17
72'	18
92'	23

Courtesy Georgia-Pacific Corp.

Fig. 103   The number of panels needed without deductions.

The beauty of real wood paneling is found in the natural variety of grain and color inherent in each panel. Before beginning application, take a few moments to place the panels around the room on each wall. Rearrange them to achieve the most pleasing balance of color and grain pattern, then number the back of each panel in sequence. Be sure to arrange the panels to achieve the groove-pattern sequence you desire, as shown in Fig. 104. The panels are random-grooved, with grooves also located on 16-inch centers. If the studs are on 16-inch centers, this means that after locating the first stud and starting at that point, you can do all the nailing through the grooved areas. The following instructions apply to installing paneling on plaster or drywall walls in good condition; applying paneling to rough plaster, concrete-block walls or concrete is covered later in the chapter.

Fig. 104   Arranging panels for installation.

STUD LOCATION

Probe through the wall with a long nail as shown in Fig. 105 to find stud location. A stud finder (Fig. 106) is also useful. The magnet in the stud finder will point to nails hidden in lath or drywall. Check at several places; tap middle of wall and bottom to determine exact location of studs. When the center of the studs has been located, snap a chalk line on the wall at center line and make a small mark on the floor and ceiling in front of the stud for use after the panel has been set in place.

CUTTING THE PANELS

Measure the floor-to-ceiling height for the first panel, starting at the corner of a room. Allow ½ inch clearance, top to bottom; molding will cover this space. If the corner is irregular, such as a

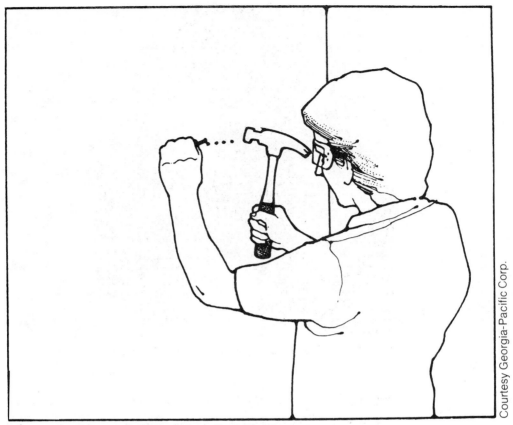

Fig. 105   Using a long nail to find studs.

brick or masonry wall, a small compass can be used to scribe the panel for perfect fitting as shown in Fig. 107. Measure panel for cutting and mark dimensions in soft lead pencil. Use a straight-edge to provide a clear, even line. If you are using a cross-cut handsaw or table saw, keep the face of the panel up, Fig. 108. Using a portable saw or a saber saw, keep facedown, as in Fig. 109.

CUTTING AROUND DOORS AND WINDOWS

Accurate measuring is important here. Measure from the last applied panel edge to the trim or to the opening if trim has been removed. Measure up from bottom and down from top to get an accurate pattern for cutout. Drill a ¾-inch hole from the panel surface side to provide a turning corner for your saw. A keyhole or saber saw is helpful for this kind of cutting. If saber saw is used, the cut should be made with the panel facedown. If possible, the cut-out panels should meet on the vertical trim line of a window or door. This should help eliminate extra cutting and large waste sections out of two panels.

Fig. 106   Using a stud finder to locate studs.

Fig. 107   Using a compass to scribe panel to fit irregular
            wall.

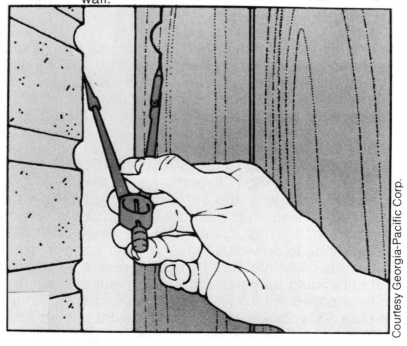

Fig. 108   When a handsaw is used, the panel
should be faceup.

Fig. 109   When a portable power saw is used, the panel
should be facedown.

OUTLET BOXES

Again, accurate measurements are important. Many a good panel-ing job has been ruined by inaccurate cutting for outlet boxes or heat registers. After the panel has been properly cut to fit, but before you make the cutout, remove the faceplate and mark the outlet box with chalk, as shown in Fig. 110. Place the panel against the wall and tap on the panel at the box location. This will transfer the image to the back of the panel and indicate the area for cutting. If the cut is made ¼ inch outside of the chalked area, the receptacle faceplate will cover the extra space around the hole. Drill ¾-inch holes through panel, then use a keyhole saw or saber saw to cut the opening.

Fig. 110   Chalked outline of box is transferred to back of panel.

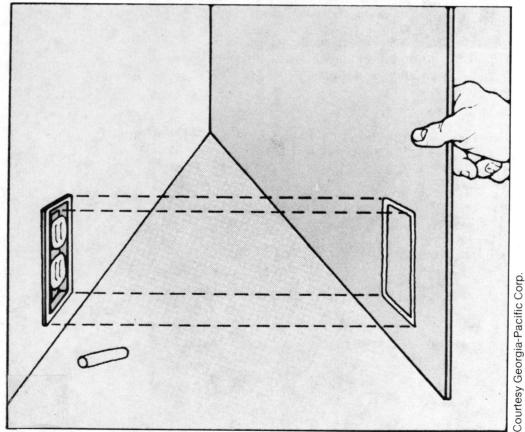

Courtesy Georgia-Pacific Corp.

USING PANEL ADHESIVE

Use of panel adhesive is the professional way to apply paneling. Follow the manufacturer's recommendations on applications direct to studs or over existing walls. Make absolutely certain panels are properly cut and fitted before applying adhesive: once adhesive is applied, it's hard to make adjustments. Check to see that walls and panels are free from dirt or particles of board, etc., before applying adhesive. The adhesive should be applied with a caulking gun, as shown in Fig. 111, to both the back of the panel and the face of the wall or studding. Use shims or blocks to keep panel snug and in place while nailing or gluing. Butt second panel into place next to the first joint, tapping it up snug with a

Fig. 111   Correct way to apply paneling adhesive.

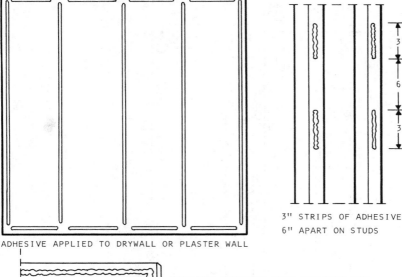

3" STRIPS OF ADHESIVE
6" APART ON STUDS

ADHESIVE APPLIED TO DRYWALL OR PLASTER WALL

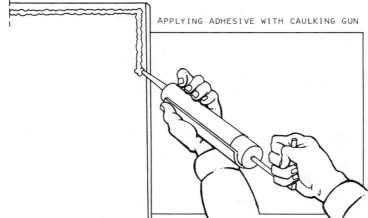

APPLYING ADHESIVE WITH CAULKING GUN

small wooden block. Regular finishing nails can be used if desired; these will require countersinking and use of a putty stick to hide them. If colored nails are used, matching the color of the grooves, countersinking and the putty stick can be eliminated. One-inch nails should be used for application direct to studs and 1⅝-inch nails through drywall or plaster.

APPLYING THE PANELS

Set first panel into place and butt it to the wall in the corner. Make sure the panel is perfectly plumb, using the level, as shown in Fig. 112, and that the outer edge is parallel with the chalked line on the center of the stud. If the outer edge does not fall directly on the

Fig. 112 Edge of panel must be plumb and on center of stud.

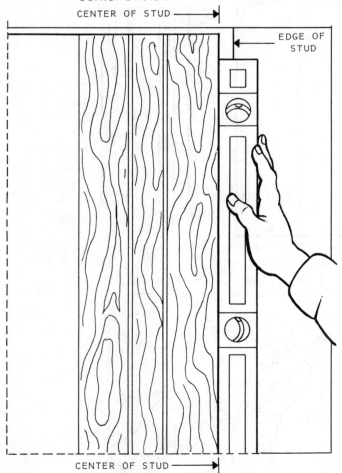

CENTER OF STUD

EDGE OF STUD

CENTER OF STUD

center of a stud, scribe or mark the panel to the adjacent wall so that the outer edge of the panel comes to the center of the stud, leaving room for nailing the next panel. When the center of the stud, the joint location, has been determined, it is helpful to paint a color stripe, as shown in Fig. 113, matching the groove color, on the wall at this location. Any noticeable gap between the panel edges will thus be minimized.

Courtesy Georgia-Pacific Corp.

Fig. 113   Painting matching stripe on wall.

PANELING UNEVEN SURFACES

Most paneling installation over existing true walls requires no preparations. Paneling over a sound level surface is quick and easy.

Basement or garage areas with poured-concrete or concrete-block walls usually require some preliminary work, after which the finished wall, straight and smooth, will make the extra work well worthwhile. The first step is to find the high spots on the wall. If the wall is reasonably straight and true, just chip off any protruding mortar or rough spots, then brush with a stiff wire brush to remove any loose chips or dust.

Furring strips can be either 1 × 2s or 1 × 4s. One-by-fours will be more split-resistant if the furring strips are to be nailed to the concrete or block wall. If the furring strips are to be nailed, "cut" nails 1½ inches long should be used. As an alternative to nailing, the furring strips can be applied to the wall with a water-proof adhesive. If the adhesive method is used, wait 24 hours after applying the furring strips before starting the paneling. If there is a noticeable high spot on the wall, start at that point with the first furring strip; then, as you work away from the first strip, use the level, as shown in Fig. 114, to check your work, using wedges where necessary to shim the strips and keep them plumb. The furring strips should clear the floor and ceiling by ½ inch, as shown in Fig. 115. The horizontal strips should be 16 inches on center, the vertical strips 48 inches on center. When the furring

Fig. 114   Using shims to plumb furring strips.

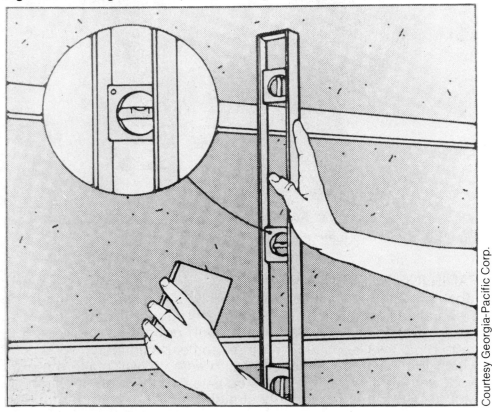

Fig. 115　Correct placing of furring strips.

strips are secured, either by nailing or with adhesive, the pre-finished paneling can be applied with nails or with panel adhesive.

MOLDINGS

Prefinished Moldings

Prefinished moldings "finish" paneling installations. They trim door and window openings to complement paneled walls, cover seams and joints at ceilings, floors, corners and other areas and protect paneling from kicks and bumps. They harmonize with paneling, have a tough surface finish that resists dirt absorption and are easily cleaned with a damp rag or sponge. Their factory-applied finish eliminates the need for finishing on the job. Easily worked and installed with common woodworking tools, wood moldings are the final step in a paneling installation. The best time to buy prefinished moldings is when the paneling is purchased, since the material requirements for both products will already be known. Also, the molding may be compared with the paneling for quick selection of the best harmonizing or contrasting tones.

Vinyl Shield Moldings

New Vinyl Shield moldings are made from polyvinyl chloride (PVC), which provides a tough, durable trim to your paneling installation, yet are as easy to work with as wood. They resist splitting, chipping and warping. Available in wood-grained colors and off-white.

Molding Installation

The secret of a good professional-looking molding job is to start with accurate measurements. Measure along the ceiling line for the cove or crown molding. Measure along the floor for exact length of base and shoe molding. Do not assume that the ceiling and floor are the same length. Fig. 116 shows the most common moldings and where they go. Your dealer can help you with specific details and installation.

Fig. 116   Names and types of moldings.

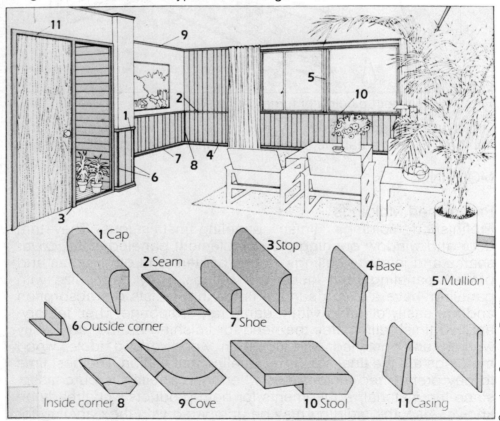

1 Cap
2 Seam
3 Stop
4 Base
5 Mullion
6 Outside corner
7 Shoe
Inside corner 8
9 Cove
10 Stool
11 Casing

Cutting: Mitering joints is the first basic operation in working with molding. For accuracy, use a miter box and fine-toothed saw. To ensure a snug fit at corners, trim both pieces at 45-degree angles in opposite cuts so together they form a tight right angle.

Splicing: To splice lengths of molding along the same wall, make 45-degree cuts at the same angles on both pieces.

Nailing: After molding has been cut properly, install with three-penny finishing nails, countersink $1/32$ inch and cover with putty stick, or use colored nails to match the molding color.

Chapter Fifteen

Suspended Ceilings

A suspended acoustical ceiling can give the basement or attic a really professional touch, and you don't have to be a professional to install it. It may take about as long to plan it as to do the actual work, but the results will be worth the trouble. You'll need to make a plan of the room, drawn to scale. You will find this an easy task if you use ⅛ or ¼-inch graph paper, available at stationery stores or drugstore school-supply counters.

Fig. 117 Ceiling plan drawn on graph paper.

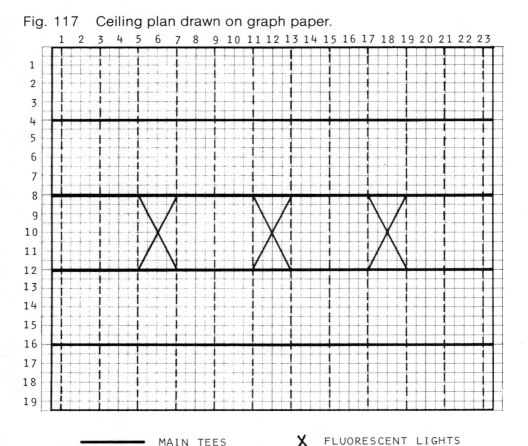

——————— MAIN TEES   X FLUORESCENT LIGHTS

- - - - - - - CROSS TEES

——————— WALL ANGLE

The grid of a suspended ceiling is made up of main tees and cross tees, shown in Fig. 118, and wall angles, shown in Fig. 119. The tees are suspended by wires from nails, hooks or screws fastened to the ceiling joists or other building members, as shown in Fig. 121.

To make the plan, draw the room size on the graph paper. In the example shown in Fig. 117, the room is 23 feet long and 19 feet wide. The standard ceiling panel is 24 inches wide by 48

Fig. 118   Main tees and cross tees support the ceiling panels.

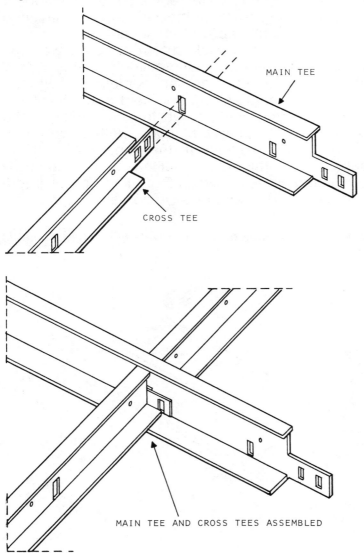

MAIN TEE

CROSS TEE

MAIN TEE AND CROSS TEES ASSEMBLED

inches long; the main tees are installed 48 inches apart, as shown in Fig. 117. Main tees should be installed at right angles to the floor joists. If recessed lighting is to be used, this should also be shown on the graph plan.

When planning a ceiling installation, everything should be worked from the center of the room; all the cut pieces should be at the borders of the room. In Fig. 117, the recessed fluorescent lights are centered and should be on individual switches for better control of the lighting. The panel grids should clear the bottom of the light fixtures by at least 6 inches. Two inches should be allowed for the grid work; thus if the basement is 8 feet from floor to bottom of the floor joists, the bottom of the wall angle (or finished ceiling) would be 7 feet 4 inches above floor level.

If the wall angle is to be attached to a concrete wall, the best type of fastener to use is the plastic shield. Shields ¼ × 1 inch, with the right-size masonry drill and sheet-metal screws, can be purchased as a kit. The angle should be held to the level line on the wall; the holes for the shields can then be marked, the angle removed and the holes drilled. Masonry drills have a tendency to drift if they hit a hard rock in the concrete, so make certain that the drill goes in straight. After drilling, drive the anchors in and install the wall angles, as shown in Fig. 119. The wall angle is shown in Fig. 120, with inside and outside corners.

Fig. 119   Wall angles can be fastened to the wall with plastic shields and screws.

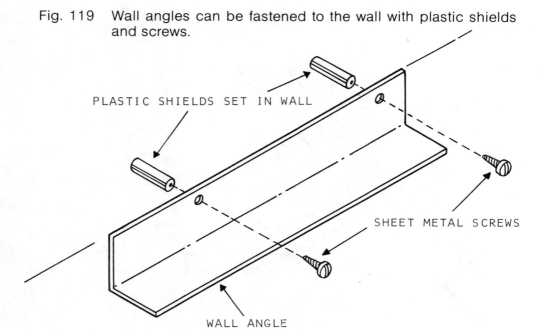

PLASTIC SHIELDS SET IN WALL

SHEET METAL SCREWS

WALL ANGLE

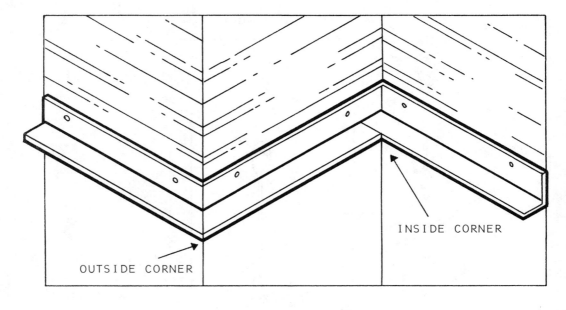

WALL ANGLE

Fig. 120   Inside and outside wall corners.

If there are breaks in the wall, such as a pilaster or pipe chase, outside corners may be needed. Outside corners should be mitered. If the wall is a straight square or rectangular room, only inside corners will be needed and they should be over-lapped, as shown in Fig. 120. The angles can be cut with tin snips or with a fine-toothed hacksaw blade.

If recessed lights are to be used, you should install them before proceeding any further.

The main tees should be installed next. The position of each main tee can be located by stretching a very tight line (nylon cord is best for this purpose) between the tops of the wall angles, as shown in Fig. 122. The suspension wires should now be cut to the proper length. The suspension wires must be long enough to reach from the top of the main tees to the nail, hook or other device to which the wire will be connected (Fig. 121). This mea-surement will depend on the methods used. The first suspension wire for each main tee should be directly above the point where the first cross tee meets the main tee. This point is found by referring back to the original sketch in Fig. 117. Attach a sus-pension wire every 4 feet along the level nylon cord guideline.

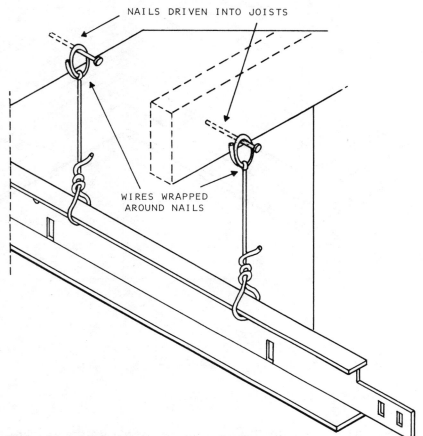

NAILS DRIVEN INTO JOISTS

WIRES WRAPPED
AROUND NAILS

Fig. 121   Main tees are supported on wires.

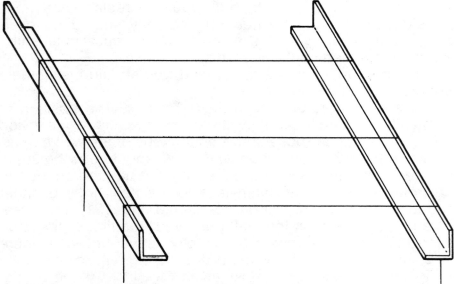

Fig. 122   A tightly stretched line is used to set main tees.

Straighten each wire and make a 90-degree bend in each wire where the wire crosses the level line, as shown in Fig. 123.

The standard measurement for the main tees is 12 feet long with cross-tee slots punched every 12 inches, beginning at 6 inches from each end. If we refer to the layout sheet, Fig. 117, we

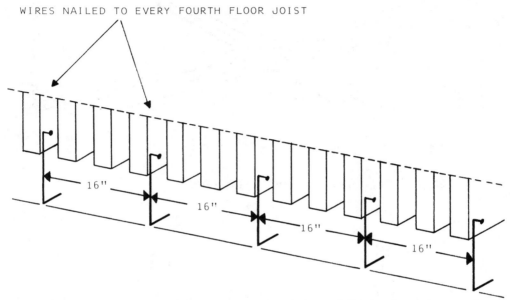

WIRES NAILED TO EVERY FOURTH FLOOR JOIST

16" 16" 16" 16"

Fig. 123 Suspension wires are bent at right height for hanging main tees.

see that the first cross tee is 6 inches from the wall. Measure this distance along the top flange of the main tee and locate the slot for the cross tee just beyond this point. Measure back from the slot the same distance minus ⅛ inch (to allow for the wall angle). Now saw the main tee at that point. The main tee and the first cross tee are shown in Fig. 124. In our example the room is 23 feet long; we will need to splice two main tees together. Whenever splices must be made, make the cuts and splices correctly so that the suspension wires will be positioned correctly. Install all the main tees, as shown in Fig. 125, and check with a level to ensure that the tees are level with the wall angles. The cross tees can now be installed by inserting the ends of the tees into the slots in the main tees, as shown in Fig. 118. The ceiling panels can be installed by tilting them, lifting them through the openings and dropping them down into place. The border panels can be cut to size.

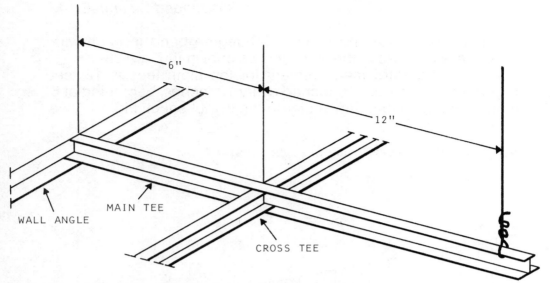

6"

12"

WALL ANGLE

MAIN TEE

CROSS TEE

Fig. 124   Locating the first cross tee.

Fig. 125   Using a level to check main tees.

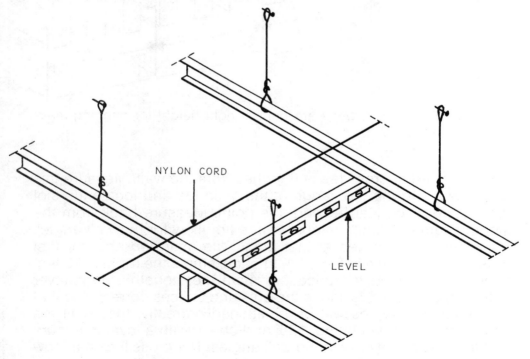

NYLON CORD

LEVEL

Lay them on a hard surface, finished side up, and use a straight-edge as a guide, cutting to the required size using a sharp knife.

Finally, again emphasizing the importance of making a layout on graph paper, the layout in Fig. 117 helps us in figuring how much material will be needed.

> 96 feet of main tees (four runs of 23 feet in 12-foot lengths)
> Sixty 4-foot cross tees
> 84 feet of wall angle (19 + 19 + 23 + 23 = 84)
> Fifty-five 2 × 4-foot ceiling panels (52 plus extras for cut pieces)
> Three 2 × 4-foot luminous panels
> Shields, screws and wire

Establishing a level line for the ceiling is simple if it's done the way the professionals do it, with a water level, as shown in Fig. 126. If soft plastic tubing or a water hose is filled with water and then held from one end of a room to the other end, the water level will indicate true level.

Fig. 126   A level line can be established using a hose, plastic tubing or a long level.

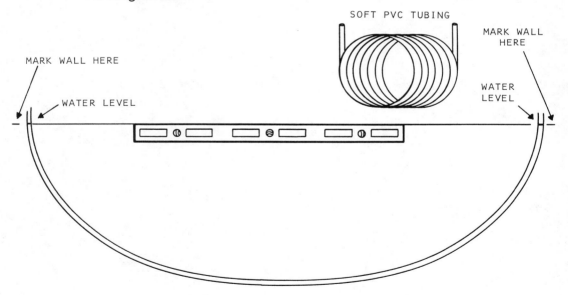

When the ceiling level has been determined at one point, one end of the tubing or hose can be held with the water level with this point, and the water level at the other end of the tube or hose can be marked on the wall. A chalk line can be snapped between the two points to establish either the top or the bottom of the wall angle, or a long level can be used. No matter how you do it, take the time to do it right. If the wall angle is level all around the room and the main tees and cross tees are level, the finished ceiling will be level.

The room size will determine the measurements of the grid work. In our example we start with a 6-inch measurement from the wall to the first cross tee; this measurement is for the example only and will vary with the room size and layout.

Acoustic panels will not be used beneath the fluorescent lights; either "egg crate" panels or frosted-plastic panels can be used below the lights.

After the gridwork is installed, all that remains to finish the job is to lay in the acoustic panels. The finish on the panels is soft and will mar or mark easily; wear clean white cotton gloves when handling the panels.

Chapter Sixteen

Ceramic Tile for Walls and Floors

There are two areas in most homes which with a little work and moderate expenditure for materials and tools can be transformed into bright cheerful rooms with easy-to-clean walls and floors. The application of ceramic tile is one of the easiest home-improvement projects a homeowner can tackle. Ceramic-tile application on walls in kitchens and bathrooms not only makes cleaning easier, it makes a room more attractive and adds to the overall value of the home. Ceramic-tile floors are beautiful and extremely durable when applied to a solid surface, preferably concrete. There will be a considerable saving in cost if you do the work; most jobs can be done for about one third of the price a contractor would charge. Only a few tools are needed and the most expensive of these, a tile cutter, can usually be rented from the tile supplier. I recommend that you do not rent this tool; buy it and you'll have it for other projects in the future. Here is a list of the tools you will need when applying ceramic tile:

1. Tile cutter
2. Tile nippers
3. Steel tape measure (6-, 8- or 10-foot)
4. Trowel with notched edges
5. Spirit level (2-, 3- or 4-foot)
6. Putty knife
7. Screwdriver
8. Grindstone to remove sharp edges on cut tile
9. Two or three small sponges for cleanup and grouting
10. Rubber gloves for grouting

There are certain basic rules you should follow when applying ceramic tile:

The walls should be clean. On old walls remove old wallpaper or vinyl covering, wash the surface and rinse well to remove soap film or grease.

Plan the design to use cut pieces, as shown in Fig. 127. The only exception to this rule is that if you are extending an

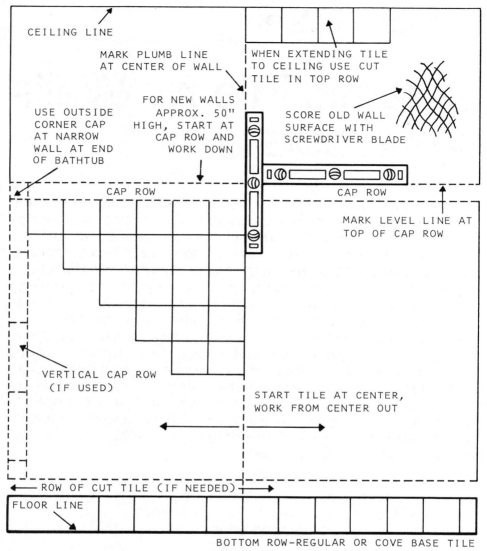

CEILING LINE

MARK PLUMB LINE
AT CENTER OF WALL

WHEN EXTENDING TILE
TO CEILING USE CUT
TILE IN TOP ROW

USE OUTSIDE
CORNER CAP
AT NARROW
WALL AT END
OF BATHTUB

FOR NEW WALLS
APPROX. 50"
HIGH, START AT
CAP ROW AND
WORK DOWN

SCORE OLD WALL
SURFACE WITH
SCREWDRIVER BLADE

CAP ROW

CAP ROW

MARK LEVEL LINE AT
TOP OF CAP ROW

VERTICAL CAP ROW
(IF USED)

START TILE AT CENTER,
WORK FROM CENTER OUT

ROW OF CUT TILE (IF NEEDED)

FLOOR LINE

BOTTOM ROW—REGULAR OR COVE BASE TILE

Fig. 127   Detail showing tile-application methods.

existing tile wall upward, such as a bathtub recess, you will almost
be forced to follow the pattern of the tile wall below. When starting
a new installation (not an extension of an existing wall), always
start at the top row or cap tile row if the tile is to extend to average
height for a bathroom (49 to 50 inches). There is no hard-and-fast
rule as to where the tile should start; plan your work so that the cut
pieces will be at the end of a row or at a corner. The work should
be planned, as in Fig. 127, working sideways from the center to a
vertical row of cut tile at the corners or end of a wall, and working
down from the cap row to a horizontal row of cut tile at or near the

bottom of the wall. Plan your work so that no cut pieces are less than 1 inch in width. When tiling around a bathtub leave ⅛ inch between the bottom of the tile and the edge of the tub; this space should be sealed with a white silicone rubber sealer.

When ceramic tile is applied to a kitchen wall, as shown in Fig. 128, every effort should be made to have the pattern work out to a symmetrical or balanced design. Wherever possible, plan the work so that all the cuts are at the end of horizontal rows. When tiling above a counter top, start at the top of the backsplash and work up, making any necessary cuts at the bottom of the cabinets. Cap tiles should be used at outside corners of walls and around recessed windows. Cap tiles are made with rounded edges, are made in both inside and outside corner types and are available in both full tile size and in 2-inch widths.

Fig. 128   Ceramic tile above counter top makes an attractive, easy-to-clean wall.

Remove the cover plates on electrical switches and receptacles when tiling a wall, and pull out the switches or receptacles approximately ⅜ inch to allow for the thickness of the tile. Cut or place the tile sufficiently close to the electrical box so that the mounting ears of the device will rest on the tile. If the original screws are too short to mount the switches or receptacles, you will need ⁶⁄₃₂ flathead screws approximately ⅜ inch longer than the original screws. The screws are standard items, available at hardware stores. The circuit breaker or the fuse controlling the kitchen switches or receptacles should be turned *off* or removed while the tiling is being done and should be left *off* until the devices and their cover plates have been replaced.

EXTENDING AN EXISTING WALL TO THE CEILING

Many builders when tiling a bathroom tub and shower recess will only apply the tile to a point 5 feet 6 inches to 6 feet above the floor. This leaves an area of approximately two to two and one half feet to be either painted or papered. Because any bathroom is at times subject to very high humidity, either paint or paper will usually show discolored areas and mold or peel eventually. There is only one satisfactory way to cure the problem: the existing tile should be extended to the ceiling, as shown in Fig. 129. Assuming a typical bathroom to be 8 feet from floor to ceiling and the original tiling stopped at 6 feet, the tile would have to be extended upward 24 inches. The length of the back wall would be approximately 58 inches and each end wall would be 29 inches (±). The wall in your bathroom will, of course, vary from this example. Ceramic tile is sold by the square foot. How many square feet will be needed to tile this area?

24 inches	58 inches	1392
×29 inches (end wall)	×24 inches (back wall)	+1392
216	232	2784
48	116	
696 square inches	1392 for a total of 2784 sq. in.	
×2		
1392 square inches		

There are 144 square inches in 1 square foot, so by dividing 2784 by 144 we know how many square feet of tile are needed.

```
        19.33, or 19⅓ square feet
144/2784.00
    144
    1344
    1296
      48 0
      43 2
       4 80
       4 32
```

Courtesy Florida Tile

Fig. 129   Extending ceramic tile to ceiling above bathtub eliminates
moisture problems.

Tile is usually sold in boxes containing 10 square feet, and since there will be some waste and breakage 20 square feet should be purchased. This may seem to be cutting it very close, but in this particular case all the cap tile now in place across the top of the existing tiled area should be saved to use again. Cap tile will be needed to extend from the top of the existing tile up to the ceiling at the two sidewalls.

In many cases, if the house is ten or more years old it may be impossible to match the existing tile color; if you should encounter this particular problem, one very practical solution is to use a contrasting color or white. The end result if white is used will be more light in the bath enclosure, and it will appear to have been designed with white tile originally.

Tile from different manufacturers may differ slightly in size; when extending an existing wall surface, if a spare piece of the original tile is not available, remove an existing tile and carry it with you to assist in matching both color and size. You will not have to worry about the spacing between tiles: they are made with raised edges (lugs) to maintain proper spacing.

The first step in extending the tiled area is to remove the row of cap tiles. This must be done carefully in order not to break them; some of these tiles will be needed to extend the vertical row of cap tiles to the ceiling after the two ends are applied. Insert a putty-knife blade under the end of the cap tile and tap the end of the knife lightly with a hammer. This will loosen the tile; it should then pull off easily. Use the putty knife to remove the old mastic which held the tile. The wall should then be scored, as shown in Fig. 127, using a screwdriver blade. After scoring the wall and before spreading the mastic, wipe the surface with a damp cloth to remove any dust. Follow the directions on the can when applying the wall mastic. Use the notched trowel to spread the mastic as directed. A gallon of mastic will usually cover from 70 to 80 square feet. If the mastic is applied too heavily, when the tile is pressed against the wall as it is set in place the excess mastic will be forced out between the tile joints, making proper grouting impossible and cleanup of the tile face more difficult. When spreading the mastic, apply it only to an area which you can cover with tile in 15 minutes. This will allow time to make any necessary cuts.

When you buy or rent a tile cutter, if you haven't used one before ask for instructions on its use. Ceramic wall tile is grooved on the back side and all cuts should be made in the direction of the grooves, not at right angles to the grooves.

The cut edges of ceramic tiles will be very sharp and often jagged; these sharp edges can be removed by rubbing the cut edges across a grindstone or coarse sandpaper. As I mentioned earlier, when laying out the tile pattern the work should be planned so that no cut pieces less than 1 inch will be used.

After spreading the mastic, set the tile into place with a slight twist and with the grooves in a horizontal position. This will help to keep the tiles from sliding down. Use the level frequently to check the vertical and horizontal alignment. After applying all the whole tile, go back and measure, cut and apply the cut tiles. Clean up any mastic on the face of the tile as you go, using a solvent or a single-edge razor blade. Water-cleanup-type mastic can be removed with a damp sponge. All wall tile should be allowed to set for 24 hours before the grout is applied.

GROUTING THE TILE

When selecting the grout try to purchase a type which uses latex liquid instead of water as a mixing agent. Latex-mixed grout is superior to water-mixed types because of its resistance to water penetration. Follow the directions on the package carefully when mixing the grout. When it is mixed to the proper consistency spread the grout on the face of the tile and, using a sponge, rub the grout into the spaces between the tile. The finished grout joint should be slightly lower than the face of the tile. An old toothbrush handle is a good tool to rake any excess grout from the joints. As the grout dries, cloudy smears will be visible on the tile. The grout should be allowed to dry for 24 hours, then polish the tile with a soft cloth. The tiled areas should not be exposed to water for 48 hours.

CERAMIC-TILE FLOORS

Would you like to replace your present vinyl, linoleum or rubber-tiled bathroom floor with a new, shining easy-to-care-for ceramic-tile floor? The same tools will be needed as for the wall tiling project. Ceramic tiling, when it is correctly installed, makes the perfect bathroom floor. The key word here is "correctly."

First of all, there is only one right way to install ceramic floor tile: *it must be laid on concrete.* Any movement or flexing of the subfloor under ceramic tile will cause the tile grouting to loosen and will cause the tile to break or to come loose. Many builders will nail down a ¾-inch wood subfloor, then on top of this, a ⅜-inch plywood or hardboard layer, in a bathroom. Then they will use tile mastic to cement a ceramic tile floor to the subfloors.

Fig. 130   Rain and snow will not damage this attractive ceramic-tile entryway.

Fig. 131   An attractive pattern in ceramic floor tile.

Wooden subfloors will flex (1) when walked upon and (2) when the wood expands and contracts as the temperature and humidity vary with seasonal changes. If you should find that your ceramic-tile floors were installed over wood, my advice is to remove the tiles when they become loose and replace them with another type of flooring. You can determine if the tile was laid on a wooden subfloor by rapping on the tile with a screwdriver handle; if the floor sounds hollow it's probably laid on a wooden subfloor.

If your home is a single-level, built on a concrete slab, you will have concrete floors in the bathrooms. As I mentioned earlier, concrete is the perfect base for ceramic tile; it will only be necessary to remove the existing vinyl or other type of floor to install the ceramic tile. If your toilet bowl is wall-mounted (bolted to the wall), it can be left in place. If the bowl is bolted to the floor it should be removed and the closet flange should be raised the thickness of the tile. This is shown in Fig. 132 and it is important. The *bottom* edge of the closet flange should be at the *top* of the finished tile floor to prevent any leakage due to an improper seal. You can do the tile work yourself, but I recommend that you call your plumber, have him take up the toilet, raise the flange and then come back

Fig. 132   Closet flange must be set correctly to prevent leaks.

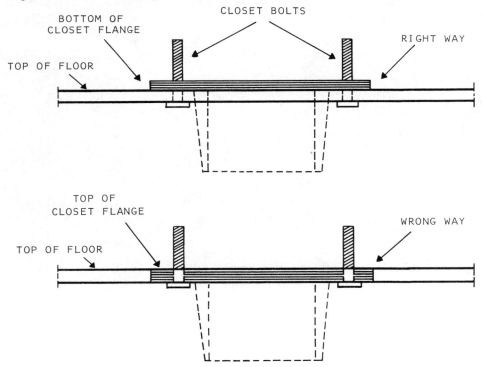

and reset the toilet after you have completed the tile installation. His services for this particular job can save you the price of a new toilet and avoid future leaks. *Don't leave the toilet in place and try to tile around the bowl!* At some future date the toilet will have to be removed and reset or possibly replaced; if the bowl is tiled in it will be almost impossible to take it up to reset it without breaking the bowl and a new bowl would not fit the opening in the tile floor.

The work of removing the old vinyl, linoleum, asphalt or rubber-tile floor can be made easier if the surface is heated. An old electric iron makes a good tool for this purpose, or the flame from a small propane torch will soften up not only the tile but also the mastic under the tile. After the tile and the mastic have been softened they can be removed with a putty knife. Care should be exercised when using a propane torch: the old materials need only to be softened, not set afire.

Determine the number of square feet of tile needed, using the formula described earlier for wall tile. Ceramic floor tile is available in many different patterns and sizes, and can be purchased mounted on a paper backing in 12-inch-square sheets (1 square foot each). Start in the approximate center of the room and apply the mastic with a serrated-edge trowel. Place the tile, with the paper backing down, on the mastic. Do not crowd the tile together; when the paper backings touch the spacing is correct. Use the tile nippers to cut the tile to fit into corners, against the bathtub apron and under and around the closet flange. It may be possible to use the existing threshold at the bathroom door by loosening it, laying the tile under it and then replacing the threshold. If a new one is needed it will be available at your tile supplier or a hardware store. When purchasing the tile and mastic, specify a mastic made for floor tile. If any mastic is smeared on the tile clean it up immediately; it will be difficult to remove after it has dried. Allow 24 hours for the mastic to set up before applying the grout.

A grout mixed with white silica sand is best for floor tile: mix 1 part of white portland cement to 3 parts of white silica sand, add water and mix to a thick soupy consistency. Apply the mixture to the tile and work it well into the joints with a sponge. The grouted joint should be slightly lower than the tile. While the grout is still wet wipe up any excess grout with the sponge. Don't worry about any cloudy smears on the tile; they will be removed when the tile is polished. Allow 24 hours for the grout to set, then polish the tile with a soft cloth and your floor is finished.

The toilet can now be reset and the floor is ready to use.

2
HOME MAINTENANCE AND REPAIRS

Chapter Seventeen

Repairing Drywall

Tools needed:

> Claw hammer (if drywall patch is to be nailed in place)
> 1½-inch-wide flexible-blade putty knife
> 6-inch-wide flexible-blade putty knife
> Drywall or keyhole saw
> Straightedge (straight strip of metal or straight board)

Material needed:

> Drywall (of proper thickness as explained later)
> Ceramic-tile adhesive
> Drywall joint compound, available ready-mixed or as a powder to be mixed with water as needed
> Drywall joint tape
> 1" × 4" lumber, short lengths

HOW TO PATCH DRYWALL

Large Holes (10 to 12 Inches Long, 6 to 8 Inches Wide)

Drywall, a gypsum-board product, is widely used for interior walls and ceilings in both commercial and residential construction. Drywall has low impact density and is easily damaged, but it does have the advantage of being easily repaired. Dents and/or scratches can be filled with drywall joint compound, allowed to dry and then can be sanded and restored to its original condition. Holes can be repaired and if the work is done carefully, the repair will be impossible to detect. Drywall is made in three thicknesses: ⅜ inch, ½ inch and ⅝ inch. When making repairs it is important that you secure the proper-thickness drywall to maintain an unbroken wall or ceiling line. Drywall is available at lumber and building-supply stores.

Store any unused drywall in a dry place for future repairs. When purchasing joint-filling compound, insist on "drywall joint compound"; some spackling compounds sold in small cans set up very hard and are difficult to sand.

As shown in Fig. 18, wall studs and ceiling joists are usually placed on 16-inch centers. If the damaged area is large enough to cover most of the space between two studs or joists, it would be best to make the repair as shown in Fig. 133. Use a straightedge to mark the area to be removed and use the drywall saw to cut out the damaged area, cutting to the *inside* edge of each 2 × 4 stud.

Fig. 133   Steps in repairing large hole in drywall.

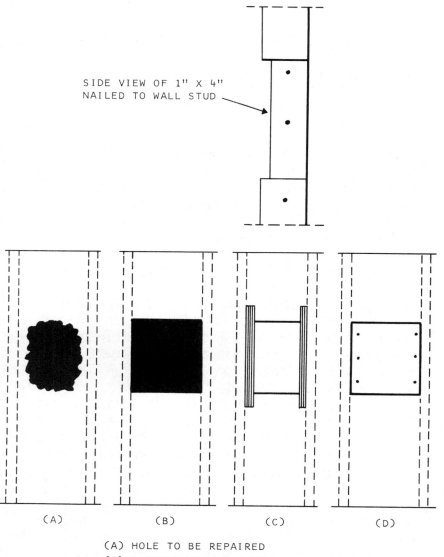

SIDE VIEW OF 1" X 4"
NAILED TO WALL STUD

(A)   (B)   (C)   (D)

(A) HOLE TO BE REPAIRED
(B) DAMAGED AREA CUT OUT
(C) 1" X 4" STRIPS NAILED TO STUDS
(D) NEW DRYWALL PATCH NAILED TO STRIPS

Measure the distance from top to bottom of the cut-out area. Cut two 1 × 4s, each approximately 8 inches longer than the top-to-bottom measurement and holding the 1 × 4s tight against the *inside* edge of the existing drywall, extending approximately 4 inches above and below the hole, nail the 1 × 4s to the studs, as shown in Fig. 133-C.

Measure the *thickness* of the existing drywall. As I mentioned earlier, it is important to secure the proper thickness. It may be necessary to purchase a full 4 × 8-foot sheet; the unused portion can be stored for future use. Measure the opening to be replaced and cut the patch a full ⅛ inch smaller, in length and width, than the cutting area. Example: If the opening is 12 × 14¾ inches, the patch should be 11⅞ × 14⅝ inches. This will allow for proper joint filling and taping. Nail the patch to the 1 × 4 strips as shown in Fig. 133-D. Use the 6-inch joint-compound knife to fill the joints with compound, apply the joint tape over the compound and smooth the tape with the knife blade. Allow this area to dry overnight, then apply more joint compound, covering the tape and extending a full inch beyond the edge of the tape. When the compound has thoroughly dried, wrap a piece of fine sandpaper around a wooden block and sand the area smooth. The repaired area should be primed with glue sizing or shellac before painting or papering.

Drywall nails are special nails used for securing drywall to wooden structure. The nails should be driven slightly below the surface of the drywall and the depression filled with the joint compound. It will be necessary to fill all joint areas and nail depressions at least three times, then sand these areas with the sandpaper and block to achieve a perfectly smooth wall surface.

Small Holes

Very small holes, nail holes and dents can be filled with joint compound. Holes 1 inch and up in diameter will require a small patch. Square up the hole to be repaired, then cut a strip of drywall 4 inches longer and 1 inch narrower than the hole. Apply a dab of ceramic-tile mastic (if you don't have some on the basement shelf, it's available in small cans at your local hardware store) to the ends of the backing strip of drywall, as shown in Fig. 134. Insert this strip through the squared hole in the drywall and pull it tight against the inside surface of the drywall. Allow 8 to 10 hours for the mastic on the backing strip to set. Cut a piece of drywall ¼ inch smaller in length and width than the squared hole and apply

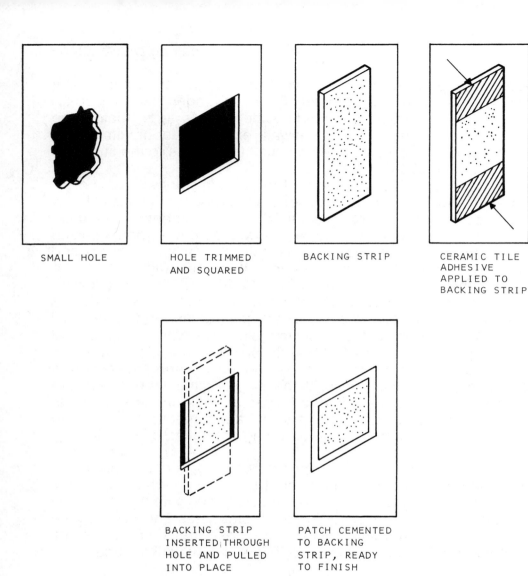

SMALL HOLE

HOLE TRIMMED
AND SQUARED

BACKING STRIP

CERAMIC TILE
ADHESIVE
APPLIED TO
BACKING STRIP

BACKING STRIP
INSERTED THROUGH
HOLE AND PULLED
INTO PLACE

PATCH CEMENTED
TO BACKING
STRIP, READY
TO FINISH

Fig. 134   Steps in repairing small hole in drywall.

a light coating of mastic to the back side of the patch and press the patch into place. After the mastic has set for at least 12 hours, the patch can be finished, using the same method as described for finishing larger holes. The final steps for repairing drywall patches are shown in Fig. 135.

Either wood strips or drywall strips can be used for the backing strips, but the drywall patch must be the same thickness as the drywall being repaired.

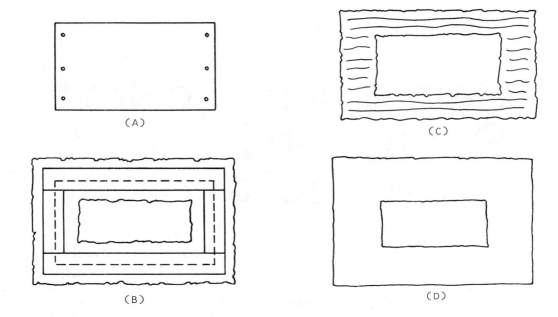

(A)   PATCH NAILED TO 1" X 4" STRIPS

(B)   NAIL DEPRESSIONS AND CRACKS FILLED
       WITH JOINT COMPOUND AND TAPED

(C)   JOINT COMPOUND APPLIED AND EDGES FEATHERED

(D)   AREA SANDED LIGHTLY, READY TO PAINT

Fig. 135   Final steps in repairing drywall.

Chapter Eighteen

Setting Wrought-Iron Railings and Anchor Bolts

Setting wrought-iron railings or anchor bolts in old concrete has always been a problem. Unless the railings or bolts were set when the original concrete was poured they soon worked loose. Melted sulfur is excellent for setting railings or bolts in old concrete, but it is difficult and dangerous to work with because of the fumes generated when the sulfur is melted.

Hydraulic cements were developed as waterproofing materials but are also excellent for use in setting railings and bolts in

Fig. 136   Setting railings and bolts in concrete.

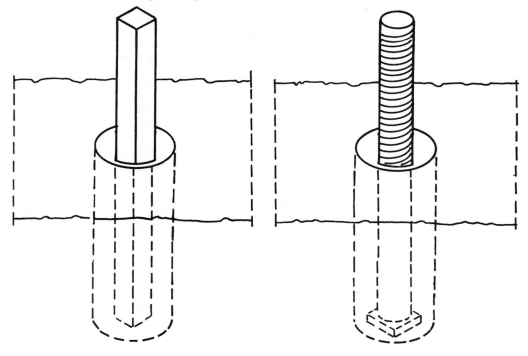

old concrete. The best types are sold in airtight cans. (Some brand names are Por-rok, Dike and Waterplug.)

The hole in concrete for a railing or bolt should be twice the diameter of the item being set; loose chips and dust should be blown or washed out and the surface of the hole should be damp but not wet. The depth of the hole will depend on the object being set, the thickness of the concrete, etc.; the holes for railings should be at least 4 inches deep if possible. Follow the directions on the can when mixing the cement; if the cement is fresh the setting-up time is usually about 1 hour.

When the cement has been mixed to the proper consistency, center the railing or bolt in the hole and pour the wet cement around it. Use a thin rod or stick to work the cement down into the hole to ensure that the wet cement goes all the way to the bottom. After the space has been filled, level the top even with the surrounding surface. When setting a railing be sure that it is plumb (vertically straight) by using a level in the same way as shown in Fig. 45. Bolts extending only a short way above the surrounding surface can be set reasonably straight and true "by eye."

Setting railings and bolts is shown in Fig. 136.

Chapter Nineteen

Easy-to-Build-and-Install Shelving

Shelving which can be used for bookcases, storage or for display purposes can be easily constructed using oak stair risers and wall-mounted brackets. The brackets and standards are available at hardware, building supply and mail order stores. The standards are available in lengths of 2, 3, 4 and 6 feet. The slots into which the brackets fit are ½ inch apart for the full length of the standard, allowing for a variation in the distance between shelves as well as variation in the height of the complete unit.

Oak stair risers are suggested for this project because oak is structurally very strong, the risers require a minimum of sanding and the oak can either be given a natural finish or stained to match the decor of the room.

Standard distance between wall studding is 16 inches center to center. Use a stud finder to find a stud approximately in the center of the area where you wish to mount the shelving. If you do not have a stud finder they are available with directions for their use, at hardware stores. After finding the center stud, as shown in Fig. 137, locate the first stud to the right and left of the center stud. The standards should be mounted to these two studs using round-head wood or metal screws 2 to 2½ inches long. The only critical measurement in this project is the exact distance between shelves when the shelves are fastened into the frame. This measurement is obtained by placing the brackets into the slots, tapping the brackets into place and measuring the *exact* distance between the tops of the two brackets. This distance will be the measurement between the bottom of the lowest shelf and the bottom of the middle shelf, as shown in Fig. 137. Three sixpenny finishing nails are all that is needed to secure the shelf to the frame. Draw a light pencil mark on the outside of the frame at the *center* of the shelf setting. Use a ¹/₁₆-inch drill bit to drill pilot holes through the frame and into the shelf. Oak is very hard wood; it will be necessary to drill pilot holes when assembling the frame also. Apply a thin coat of casein-type carpenter's glue to the end of the shelf. Line the shelf

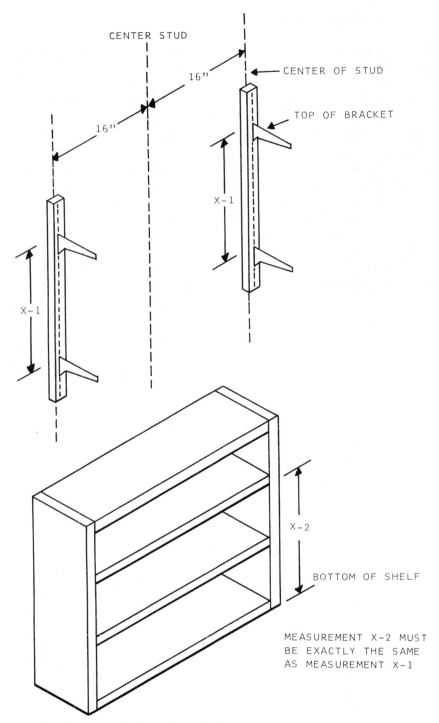

CENTER STUD

CENTER OF STUD

TOP OF BRACKET

16"

16"

X - 1

X - 1

X - 2

BOTTOM OF SHELF

MEASUREMENT X-2 MUST
BE EXACTLY THE SAME
AS MEASUREMENT X-1

Fig. 137   Hanging wall shelving.

up with the pilot holes and nail the shelf to the frame. Wipe off any excess glue immediately with a damp cloth. When the glue has dried the shelving can be finished with a coat of clear shellac and a coat of varnish for a very attractive golden oak finish, or they can be stained to a darker color and varnished if desired.

When mounting the standards on the wall, measure up from the floor to the desired height for the bottom of the standards and hold each standard to this point. Assuming the floor to be level, this will ensure that the finished shelving will be level. If the floor is not level, measure up to the bottom of one standard and use a spirit level to draw a level line over to the point where the other standard is to be mounted.

After nailing and gluing the shelving and before using shellac or varnish, set the nails and fill the nail holes with putty or plastic wood. When the shelving is finished, the nail holes will be almost invisible.

Chapter Twenty

Replacing Window Glass

At one time or another all of us are confronted by the task of replacing a broken pane of glass in a door or window. There are four principal ways by which glass is held in a frame:

1. By tack points and putty
2. By tack points and wood molding strips
3. By spring clips, with or without putty
4. By plastic snap-in strips

If the glass was secured by tack points and putty and the putty was painted to match the frame, putty or glazing compound should be used when the glass is replaced. If desired, the putty or glazing compound can then be painted to match the frame. Save the tack points when removing the broken glass; they can be used again when replacing the glass. Tack points can be pressed into place using the flat side of a screwdriver blade. When the pane of glass was held in place by wood molding strips, the strips should be removed by inserting a thin instrument such as a putty knife, knife blade or small screwdriver between the strips and the frame and prying the molding away from the frame. Spring clips are generally used only on basement metal frame windows and the glass should always be sealed with putty, glazing compound or silicone sealer to stop air and water infiltration. The plastic strips which often secure the newer type of window glass become brittle with age and will usually break or split when they are removed. If the strips are damaged when removed, discard them and use silicone sealer to seal the pane in place.

Measuring the glass. No matter how the original pane was held in place, when the broken glass has been removed, the recessed area to receive the new pane will be exposed. This is shown in Fig. 138 as 24 inches long and 18 inches wide. Allow ⅛ inch less, both

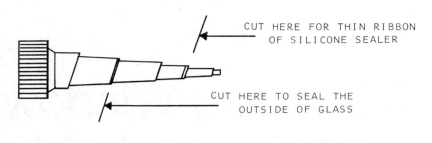

CUT HERE FOR THIN RIBBON
OF SILICONE SEALER

CUT HERE TO SEAL THE
OUTSIDE OF GLASS

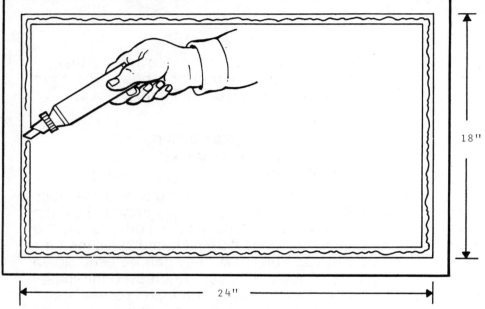

18"

24"

Fig. 138   Installing window glass in frame.

in length and in width, when ordering the replacement pane. This will allow $1/16$-inch on each side—more than enough for easy insertion into the opening.

Sealing the glass. If putty or glazing compound is used, apply a small amount of either to the recessed area to bed the glass in. When the glass is set in place, apply pressure lightly to the glass near the edges to bed it evenly. If tack points are used, press them into place with a screwdriver blade. If wood molding strips are used, replace them; if putty or glazing compound is used, knead the putty between your hands, working excess oil into it until it becomes soft and pliable, then use a putty knife to smooth it into place. If plastic strips were used originally and they were damaged when removed, silicone sealer makes an excellent substitute. As shown in Fig. 138, a thin ribbon of silicone sealer is applied to the recessed area. The pane of glass can then be set in place and the sealer should be applied around the outside of the glass to make a water- and airtight seal. The applicator nozzle should be cut to cut number 2, as shown in Fig. 138, for sealing the outside. With just a little practice you can learn to bevel the edge of the silicone to just the right angle as you apply it. Silicone sealer is made in clear (colorless) and white types. It is sold under several trade names: GE silicone sealer and Chatham Plumbers silicone seal are two; either of these is excellent for replacing glass. They are sold card-mounted, with applicator nozzle; the white type should be used for glass replacement.

Chapter Twenty-One

Patching Broken or Cracked Concrete

Corners or edges of concrete steps often break, due to weather or dropped objects. Concrete often cracks, usually due to settlement of the ground under the concrete. Broken steps can be repaired; one method is shown in Fig. 139.

The first step is to remove all loose material from the area to be repaired. Two boards will be needed to form the corner. These boards should extend at least 1 inch beyond the broken area and at least 1 inch above the top of the step. One method of bracing these forms is shown in Fig. 139; another way is to set a concrete block against each board.

The best concrete mixture for step repair is a 3-2-3 mixture, that is, 3 parts of sand, 2 parts of pea gravel, 3 parts of portland cement. This mixture is very rich in portland cement, which will add strength. After cleaning all loose material from the area to be patched, soak the area thoroughly with water. A slurry mixture of equal parts of sand and portland cement and enough water to mix to a thick soupy consistency, should be applied to the area to be patched. A paint or whitewash brush makes a good slurry applicator. Next, mix the concrete, adding only enough water to allow the concrete to flow into the patched area. Use a stick or metal rod to "puddle" the mixture, as shown in Fig. 140, after filling. Puddling will ensure that the concrete mixture flows into all the niches and rough areas, bonding the patch to the original material and eliminating voids (honeycomb) in the finished patch. Smooth the top with a small trowel; this will work the pea gravel down below the surface and bring the sand-and-cement mixture to the top. When the concrete has begun to set (harden), trowel the surface again, then use a broom to roughen the surface slightly. Leave the form boards in place for at least 12 hours. After removing the forms, cover the patch with burlap, rags or newspapers and keep the area wet for 24 hours to permit the concrete to cure slowly.

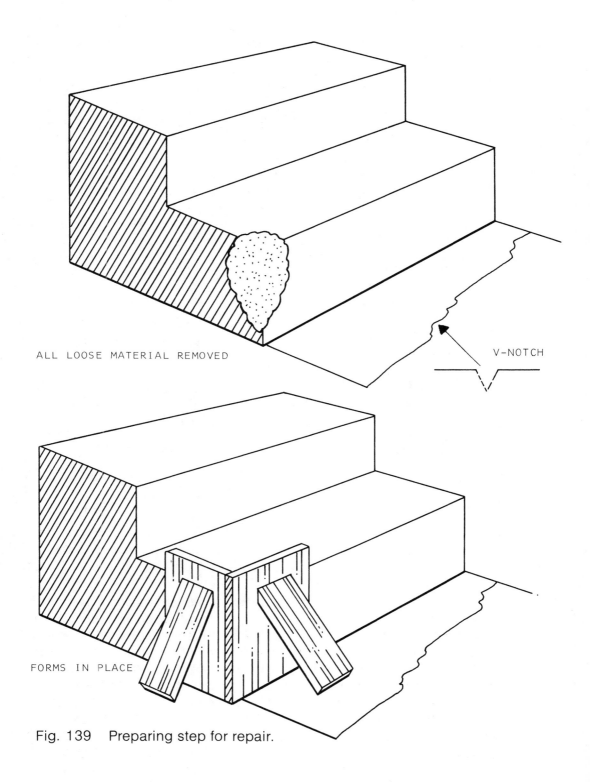

ALL LOOSE MATERIAL REMOVED

V-NOTCH

FORMS IN PLACE

Fig. 139   Preparing step for repair.

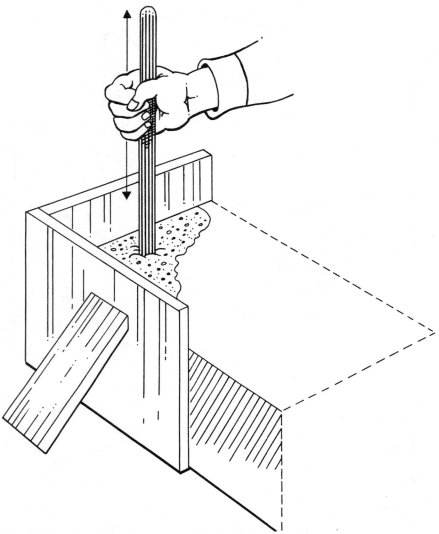

Fig. 140   Puddling wet concrete in form.

      Cracks in concrete can be repaired by using a hammer and cold chisel to open up the crack to a V-shaped notch, as shown in Fig. 139. All loose material should be removed from the crack and the area should be wet thoroughly. It should then be painted with the cement slurry mixture as described earlier. Sand-mix concrete is available at building-supply and hardware stores. Add water to the sand-mix concrete; add only enough water to make a stiff mixture. Fill the V-shaped notch with this mixture and work the mixture down into the cracked area; when the crack is full, trowel the area smooth. When the concrete has started to "set," cover the patched area with wet rags or burlap; keep the area wet for 24 hours to allow the concrete to set properly.

Chapter Twenty-Two

Adding a New Surface to Cracked or Pitted Porches

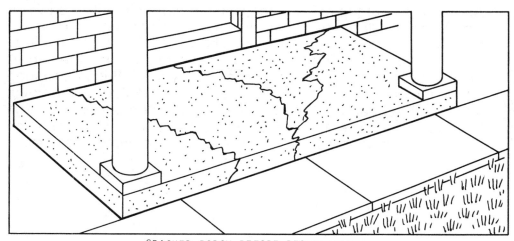

CRACKED PORCH BEFORE RESURFACING

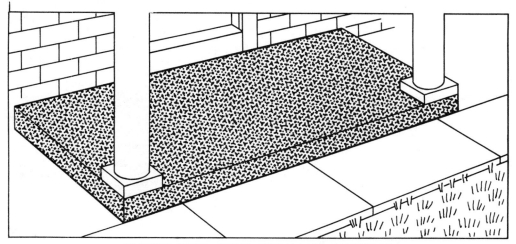

CRACKED PORCH AFTER INSTALLING CARPETING

Fig. 141   Indoor-outdoor carpet makes excellent porch repair.

What can you do when a concrete porch or patio settles and cracks, or flakes and crumbles? One quick, easy and very attractive repair is to cover the whole damaged area with indoor-outdoor carpet. There are several advantages to this type of repair: it is inexpensive, it will add color and will provide a nonskid surface to what may have been a very slick surface when wet or frozen. The concrete surface should be prepared by removing any loose material and the cracks should be filled with a cement slurry. The slurry is made by mixing equal parts of sand and portland cement, adding water to make a thick soupy mixture. A sponge or an old paint brush makes a good applicator. Pour the slurry over the cracks and work it well into the cracks with the brush or sponge. Allow this mixture to dry before installing carpet. Fig. 141 shows a typical porch slab which has cracked. The cracks could be repaired with a slurry mixture but, as shown in Fig. 141, if the slurry mixture is used first and then the porch is covered with the carpet, the finished surface will be smooth and unbroken. When measuring for the amount of carpet needed, be sure to include the vertical sides. Cut the carpet at the corners and fold the edges over. The floor-covering store that sells the carpet can also furnish the proper type of adhesive and the instructions for installing carpet.

Chapter Twenty-Three

Cleaning Roof Gutters and Downspouts

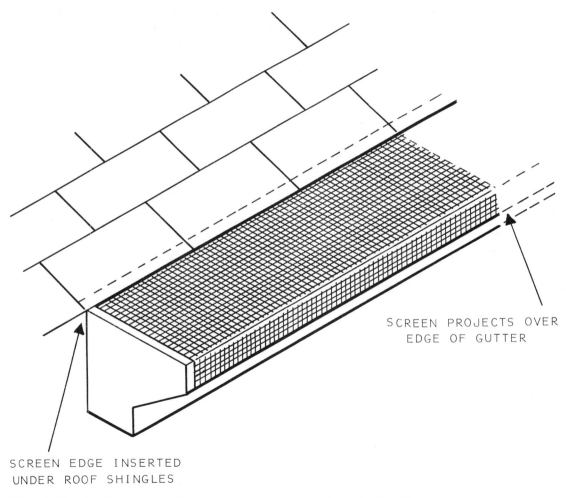

SCREEN PROJECTS OVER
EDGE OF GUTTER

SCREEN EDGE INSERTED
UNDER ROOF SHINGLES

Fig. 142   Gutter screen keeps leaves and seeds out of gutter.

If you live close to a wooded area or have trees in your yard which give birth to literally thousands of seeds each spring, you should either screen your gutters, as shown in Fig. 142, or plan on cleaning the gutters twice a year, once in the spring and again in the fall. In the spring, seeds falling by the thousands from nearby trees will land on roofs, wash into the roof gutters during a rain and then block the gutter at the downspout opening or at a corner of the gutter.

In the fall, leaves from these same trees may fall into the gutter in such numbers that the gutter and downspout are completely blocked. If this should happen and go unnoticed, water trapped in the gutter or downspout could freeze during the winter, causing severe damage and requiring expensive repair or replacement. Quarter-inch galvanized hardware cloth is excellent gutter-screening material.

Chapter Twenty-Four

Roof Leaks

Tools and material needed for this project:

> Ladder to get on roof
> Putty knife
> Asphalt-asbestos roof cement

I haven't specified how much asphalt roof cement might be needed. It's a good idea to keep a gallon (5 pounds) of this material on your paint shelf. It doesn't spoil and you'll have it when you need it.

Water leaks through a roof usually occur around the chimney flashing or the plumbing vent pipe flashings. If the roofing generally is in good condition, any leak other than at flashings, will probably be due to shingles being blown away or torn loose in a windstorm. Missing or damaged shingles should be replaced immediately.

Fig. 143 Chimney flashings should be sealed at the points shown.

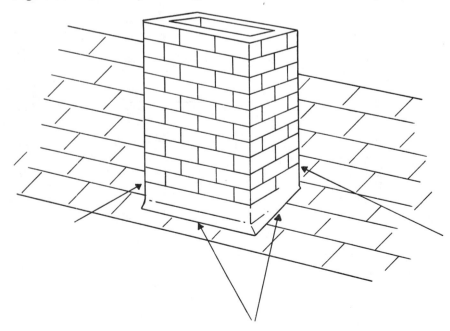

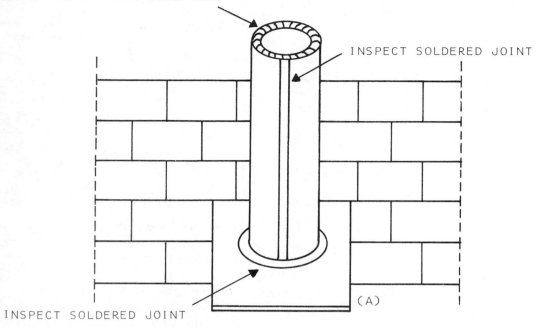

CHECK FOR A TIGHT SEAL HERE

INSPECT SOLDERED JOINT

INSPECT SOLDERED JOINT

(A)

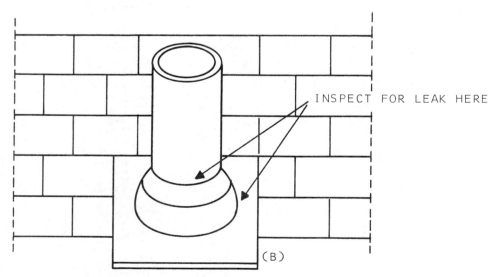

INSPECT FOR LEAK HERE

(B)

Fig. 144   Two types of roof flashing.

Leaks around flashings at chimneys are usually due to separation of the metal flashing from the chimney or the roof. This will leave a gap where water can enter. In Fig. 143, we see a typical chimney. A front-and-side view is shown. The flashing continues across the back or high side and also on the side not visible. If leaks occur at the chimney, roof cement should be applied to the top of the flashing where it meets the roofing. Two types of vent flashings are shown in Fig. 144. Type A is a lead boot flashing. This type has two soldered joints: one vertical, forming the tube, and one where the tube joins the base. After the flashing is made it is slipped down over the vent pipe with about 1 inch of lead extending above the vent pipe. A hammer is used to turn the lead down into the vent pipe and tight against the pipe. Properly installed, this type of flashing will last as long as the house, but if the very top of the lead was carelessly hit while it was being tapped into place, a future leak could be caused. If you find a leak, apply roof cement to the top, the vertical solder joint and the soldered joint at the base. Nails should not have been used with this type of flashing; if any nails are visible, cover them with roofing cement. While you're on the roof, if you have any doubts about whether water could be getting under the shingles at the sides of the flashing, apply cement to the base where it meets the roofing as a little extra insurance. Another type of flashing is shown in Fig. 144-B. When this type is used, the vent pipe projects through the flashing and the joint is sealed by a lead ring or a rubber slip joint. Apply roof cement to the top of the lead ring or rubber slip joint, as shown by the arrow in Fig. 144-B, to stop leaks at this type of flashing.

When you're on the roof repairing leaks, be generous with the roof cement: a little extra time and a little extra cement can prevent damage to your home and the trouble of climbing back onto the roof again.

Chapter Twenty-Five

Finding and Repairing Leaks Around Bathtubs

Leaks often develop around a bathtub or shower and become visible as a wet spot on the ceiling below. There should be an access panel on the wall at the end of the tub where the faucets and drain are located. It may be in a linen closet or in a hallway or another room, depending on the room arrangement. Remove the access panel and inspect the piping for signs of a leak. If no leakage is visible, turn on both the hot and the cold water and check again. If there is still no sign of a leak, and with the water still turned on, have someone direct the flow of water directly onto the wall at the faucet handles while you are watching the area behind the access panel. If the leak is at the faucets or the spout, it should be visible. A leak here can be stopped by sealing around the spout and the escutcheons, as shown in Fig. 145.

The best sealant to use is either a white or clear silicone sealer. (GE silicone sealer and Chatham Plumbers white silicone sealer are best for this.) The sealer must be a silicone sealer—other types will not work—and the area where the sealer is applied must be thoroughly dry. If the leak is at a packing nut on a shut-off valve in the access space, tighten the packing nuts ⅛ or ¼ turn, clockwise.

If there is no access panel, try sealing the areas shown in Fig. 145; if the leak persists it may be necessary to cut into the wall where the access panel should have been installed. After the wall has been cut, an access panel can be made from plywood and installed over the hole.

If water leaks from under a shower door (and it may be necessary to check this while the shower is being used), a new wipe strip may be needed at the bottom of the door. A company that sells shower doors should be able to furnish a new wipe strip; I suggest that you take the old one with you when trying to find a new one in order to match it as closely as possible. The wipe strip is secured to the bottom of the shower door with metal screws.

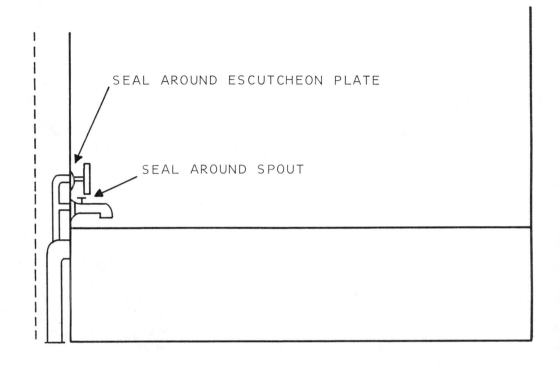

SEAL AROUND ESCUTCHEON PLATE

SEAL AROUND SPOUT

BEHIND ACCESS PANEL AT END OF BATHTUB

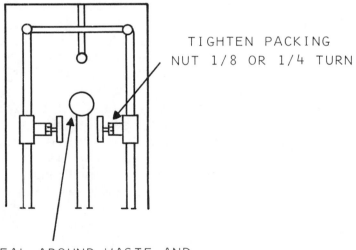

TIGHTEN PACKING
NUT 1/8 OR 1/4 TURN

SEAL AROUND WASTE AND
OVERFLOW CONNECTION TO TUB

Fig. 145   Some likely sources of leaks around bathtubs.

Chapter Twenty-Six

Setting and Connecting a Toilet Combination

Closet bowls are secured to the floor by closet bolts. These are made in two sizes, ¼ inch and $5/16$ inch; $5/16$ inch is the preferred size. A wax ring is used to provide an airtight and watertight seal between the bowl and the closet flange. The first step in setting the bowl is to insert the closet bolts under the closet flange, either under the notches at the sides of the flange or into the slots at the sides. Turn the bowl upside down and push the wax ring onto the horn, or discharge opening, of the bowl. Turn the bowl over and set it onto the flange; the closet bolts will project through the holes at the sides of the bowl. Drop the chrome-plated washers over the bolts and start the open nuts on the bolts. Tighten the nuts *hand-tight only* at this time.

Instructions for connecting the tank to the bowl should be furnished with the tank. The tank may be mounted to the bowl by two, three or four bolts, depending on the manufacturer. A soft rubber washer is used between the tank and the bowl. The washer fits over the flush valve at the bottom of the tank. Rubber washers are also used under the heads of the bolts securing the tank to the bowl. When tightening these bolts, tighten each one a little at a time; do *not* overtighten them. Keep the tank level as you work. If the tank touches the bowl before you think the bolts are tight, *stop*. There is no give in a porcelain fixture; overtightening the bolts will break the bowl. To finish tightening the bowl to the flange, use a small (6- or 8-inch) adjustable wrench and, alternating between the bolts, tighten them only until the bowl no longer moves easily from side to side on the floor. If necessary you can tighten the bolts again in a day or two. Toilet bowls can be broken very easily if the bolts are overtightened. When tightening is completed, the excess threads projecting through the open nuts can be sawed off using a fine-toothed hacksaw blade.

When connecting the water supply to the tank, if a flexible type of closet supply is used, it can be bent and shaped to fit if necessary.

A typical closet combination is shown in Fig. 146.

Fig. 146 How to set and connect toilet.

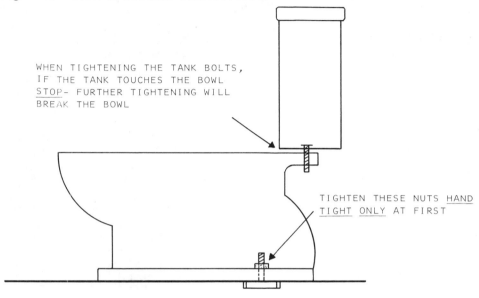

WHEN TIGHTENING THE TANK BOLTS,
IF THE TANK TOUCHES THE BOWL
STOP- FURTHER TIGHTENING WILL
BREAK THE BOWL

TIGHTEN THESE NUTS HAND
TIGHT ONLY AT FIRST

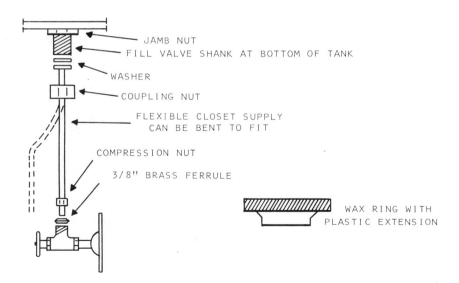

JAMB NUT

FILL VALVE SHANK AT BOTTOM OF TANK

WASHER

COUPLING NUT

FLEXIBLE CLOSET SUPPLY
CAN BE BENT TO FIT

COMPRESSION NUT

3/8" BRASS FERRULE

WAX RING WITH
PLASTIC EXTENSION

Chapter Twenty-Seven

Setting and Connecting a Self-Rimming Lavatory

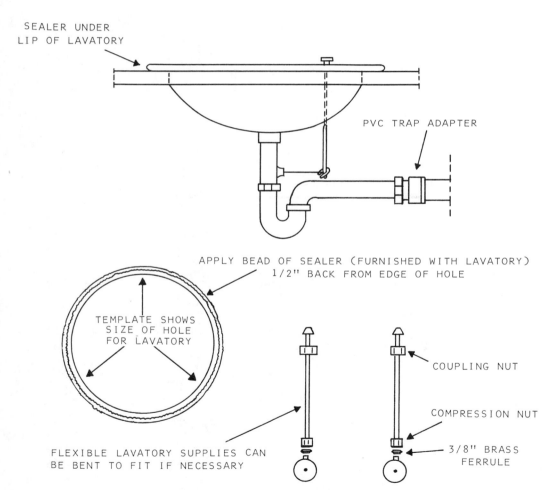

SEALER UNDER
LIP OF LAVATORY

PVC TRAP ADAPTER

APPLY BEAD OF SEALER (FURNISHED WITH LAVATORY)
1/2" BACK FROM EDGE OF HOLE

TEMPLATE SHOWS
SIZE OF HOLE
FOR LAVATORY

COUPLING NUT

COMPRESSION NUT

3/8" BRASS
FERRULE

FLEXIBLE LAVATORY SUPPLIES CAN
BE BENT TO FIT IF NECESSARY

Fig. 147   Setting and connecting a self-rimming lavatory.

The lavatory shown in Fig. 147 is a self-rimming type. It requires no metal rim or clamps to secure it in place. A template should be furnished with the lavatory, and the hole in the cabinet top should be cut to the size of the template. A tube of adhesive caulking should be packed with the lavatory; this sealer should be applied just inside the hole in the cabinet top so that the rim of the lavatory will be setting in the adhesive. If, as sometimes happens, the adhesive is missing, GE white silicone sealer is a good setting agent. If possible, spread the sealer, set the lavatory and wait at least 2 hours to allow the sealer to set up before making the water and waste connections. The faucet and the drain connection should be mounted on the lavatory before it is set in place. A ring of soft putty should be placed around the strainer on the inside of the lavatory when the drain is assembled. If flexible (chrome-plated soft copper tubing) lavatory supplies are used for the water connections, they can be bent to fit as needed.

The trap is inserted into the trap adapter, as shown in Fig. 147.

Chapter Twenty-Eight

Replacing the Working Parts in a Toilet Tank

At some period in the life of every water closet tank there comes a time when some or all of the inner workings must be repaired or replaced. Recent design changes and new materials make replacement, rather than repair, of these parts most practical. Old-fashioned ball cocks with float arms and float balls can in most cases be replaced with a new type of fill valve requiring no arm or ball. Installation is quick and simple, and only one tool is required: large slip-joint pliers. The water level can be adjusted by sliding a clip up or down on the body of the valve. I do not recommend this type of valve for one-piece toilets or in tanks using large tip-type flush valves.

The flexible refill tube may interfere with the action of the tip-type flush valve with an integral overflow tube. If the tank has a small tilt-type flush valve (American Standard), the new fill valve must be positioned to clear the flush valve when it is fully tilted. To install the fill valve, first shut off the water supply to the tank. Trip the flush lever, then sponge out the water remaining in the tank. Grasp the coupling nut at the bottom of the ball cock, and turn it clockwise to loosen and remove it. Grasp the jamb nut, up next to the tank, turn it clockwise viewed from the top to loosen and remove it. The old ball cock can now be lifted up and out of the tank.

When assembling the new fill valve, follow the instructions on the box it came in. Slide the flexible refill tube onto the fill valve before inserting the valve in the tank. When the fill valve is positioned in the tank, start the new jamb nut on the fill-valve shank, and holding the fill valve in position with one hand, tighten the jamb nut to hold the fill valve securely in place. Several different types of washers are included in the fill-valve package. If one is needed, insert it between the shank and the supply tube, then start the coupling nut on the shank and tighten the coupling nut. When

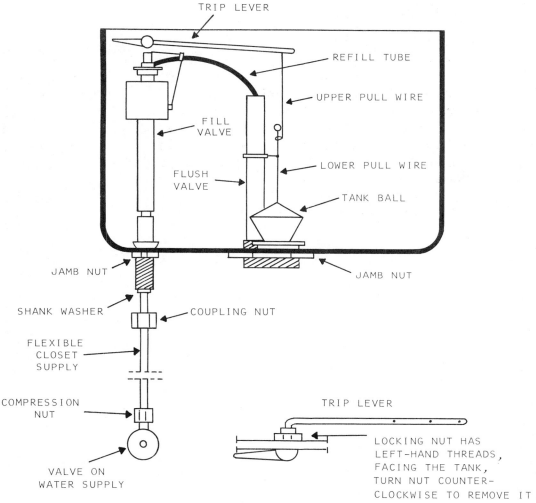

TRIP LEVER

REFILL TUBE

UPPER PULL WIRE

FILL VALVE

FLUSH VALVE

LOWER PULL WIRE

TANK BALL

JAMB NUT

JAMB NUT

SHANK WASHER

COUPLING NUT

FLEXIBLE CLOSET SUPPLY

COMPRESSION NUT

TRIP LEVER

LOCKING NUT HAS LEFT-HAND THREADS, FACING THE TANK, TURN NUT COUNTER-CLOCKWISE TO REMOVE IT

VALVE ON WATER SUPPLY

Fig. 148   Names of parts and connections for closet tanks.

starting the coupling nut be sure it has started straight, not cross-threaded. If the flexible supply tube connecting to the fill valve was moved during the process of installing the fill valve, it may be necessary to tighten the nut connecting the supply tube to the shutoff valve to prevent any leakage at this point.

Tank balls deteriorate over a period of time and become useless. A ball that has rotted will stain your hand black when you touch it. Hold the lower pull wire with one hand and turn the tank ball clockwise to remove it. If the small brass nut embedded in the tank ball stays on the pull wire, hold the wire and unscrew the nut. Then replace the tank ball. The pull wires must be straight in order to slide smoothly through the guide. Worn tank balls will permit water to leak from the tank, often resulting in high water bills. If water continues to leak after the tank ball is replaced, it may be necessary to replace the flush valve, and this is a job for a plumber. The trip lever is shown in Fig. 148 and can be replaced by turning the nut on the inside of the tank counterclockwise (viewed from the front of the tank) to remove it.

Both the old trip lever and the new replacement will have left-hand threads. The fill valve shown in Fig. 148 is a Fluidmaster 400, which has an antisiphon device to comply with plumbing codes.

If the soft-plastic refill tube is too long it can be cut to the proper length with scissors or a pocket knife.

Chapter Twenty-Nine

One Reason for Low Hot-Water Pressure

It is very common for homeowners to begin to notice that water pressure at the hot-water faucets is not as good as it is at cold-water faucets. This condition tends to build up slowly, and in almost all cases is due to lime or mineral buildup at the point where the relief valve is installed in a tee at the hot-water discharge from the heater. This condition will not occur in newer types of heaters which have a separate opening for the relief valve. As you can see in Fig. 149, the opening in the heater and through the tee is partially blocked by the relief valve sensing element. Then over a period of time lime and scale are precipitated out of the water as it is heated and the lime and scale build up at the hot-water outlet until finally there is a partial or complete blockage. To cure this condition, first shut off the cold-water supply to the heater. Open a hot-water faucet to relieve the pressure. Then remove the old relief valve. Two pipe wrenches will be needed—one to hold the fitting or tee into which the relief valve is screwed, the other to unscrew the relief valve. The relief valve must be turned counterclockwise to remove it; the proper way to hold two wrenches, using one as a backup wrench, is shown in Fig. 150. After the relief valve has been removed, insert a large screwdriver into the fitting. Use the screwdriver blade to break the lime and scale deposits loose. The loosened material will drop into the tank and settle to the bottom.

Just as scale and lime deposits form in the outlet from the tank, they will also have formed in the body of the relief valve. A new temperature and pressure relief valve of the same rating as the one that was removed should be installed. A tag mounted at the top of the old valve will show the temperature and pressure rating.

When installing the new valve, use either pipe joint cement or tape-type thread lubricant on the male relief valve threads.

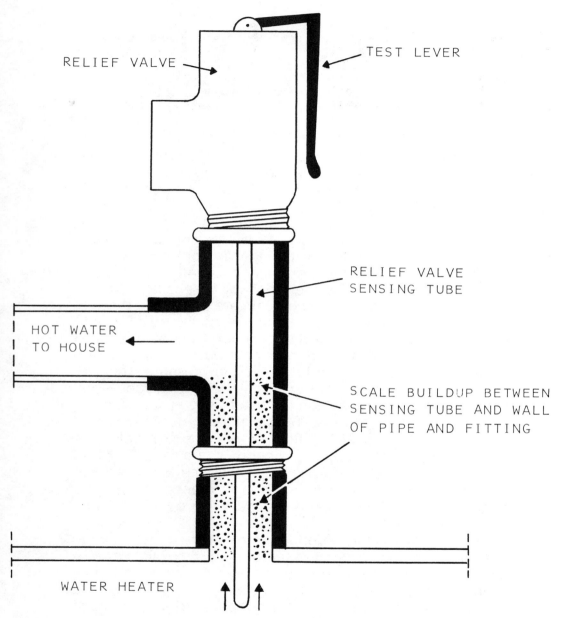

RELIEF VALVE

TEST LEVER

RELIEF VALVE
SENSING TUBE

HOT WATER
TO HOUSE

SCALE BUILDUP BETWEEN
SENSING TUBE AND WALL
OF PIPE AND FITTING

WATER HEATER

Fig. 149   Lime and scale buildup between sensing tube and walls of
pipe and fittings restricts hot-water flow.

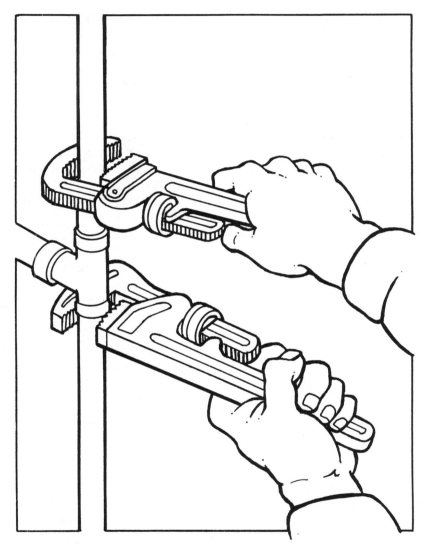

Fig. 150   Always use two wrenches, one as a backup wrench, when removing a piece of pipe or a fitting.

Chapter Thirty

Circulating Hot-Water Piping

A circulating hot-water line, correctly installed, provides hot water at the turn of a faucet.

When the kitchen or bathrooms are located at the far end of the house, or upstairs, away from the water heater, it is usually necessary to open a hot-water faucet and let quite a bit of water run before the water gets hot. The reason for this is that the water lying in the piping between the water heater and the faucet has cooled off and must be replaced by hot water from the heater. This results in waste of water.

This problem can be prevented in new homes and cured in many older homes which have a basement or crawl space by installing a circulating line. A circulating line is a continuation of the hot-water piping, extending from the farthest fixture connection, back to an extension of the drain connection on the hot-water heater.

The system operates on the principle of convection: hot water is lighter or less dense than cold water, and as the water heats in the tank it moves upward. As it moves it pushes the water in the piping away from the heater. As the water moves away from the heater, it cools and becomes heavier, and thus starts circulating back toward the heater. Fig. 151 shows the basic principle of a hot water circulating system.

Fig. 152 shows how a circulating line can be installed in a new home or in an existing home with basement or crawl space, using a circulating pump. Circulating lines will operate by gravity if properly installed, with the hot-water piping sloping upward away from the heater, and the return line (the piping from the end of the hot-water piping back to the heater) sloping down toward the heater.

As an alternate to gravity circulation I recommend the installation of a small circulating pump, controlled by an aquastat (water thermometer and electrical switch). The pump should be installed on the return line close to the water heater. The pump is inexpensive in initial cost and in operation.

The aquastat has an adjustable setting; when the water temperature drops below the preset level, the pump turns on automatically and runs until the water has reached the preset level.

Pumps using a built-in aquastat are also available.

Fig. 151   How a gravity hot-water circulating line works.

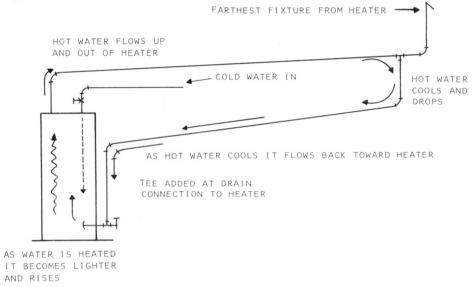

FARTHEST FIXTURE FROM HEATER ⟶

HOT WATER FLOWS UP
AND OUT OF HEATER

COLD WATER IN

HOT WATER
COOLS AND
DROPS

AS HOT WATER COOLS IT FLOWS BACK TOWARD HEATER

TEE ADDED AT DRAIN
CONNECTION TO HEATER

AS WATER IS HEATED
IT BECOMES LIGHTER
AND RISES

Fig. 152   Circulating hot-water piping system using a pump and aquastat to maintain circulation.

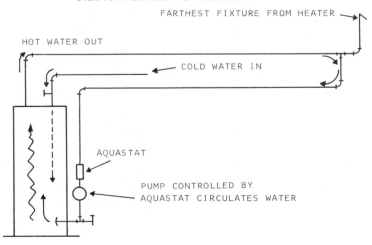

FARTHEST FIXTURE FROM HEATER

HOT WATER OUT

COLD WATER IN

AQUASTAT

PUMP CONTROLLED BY
AQUASTAT CIRCULATES WATER

Chapter Thirty-One

Soldering
Copper Tubing

Some do-it-yourselfers who can do anything that comes along seem to have trouble making soldered joints in copper tubing. If you will follow the instructions I give you, I will guarantee that you will make perfect solder joints. There are certain rules you must follow:

1. The tubing and fittings must be dry. Water will cool the copper, keeping it from getting hot enough for solder to flow.
2. The tubing end and the socket end of the fitting must be cleaned. Solder will not flow onto and stick to dirty copper.
3. Solder paste should be applied to the tubing end and the socket end of the fitting *after* both are cleaned.
4. The flame from the torch must be played on the pipe and fitting, as shown in Fig. 153, playing the flame all the way around the *back* of the fitting socket to draw the solder into the joint.
5. The fitting must be heated hot enough for the solder to flow, but not overheated. When the tip of the solder is touched to the hot fitting and the solder melts and disappears (into the fitting) the pipe and fitting are hot enough. Draw the end of the solder all the way around the fitting and stop! You've made a good joint. Do not move the freshly soldered joint for a few seconds; let the solder set.
6. Use only *plain* 50/50 wire solder. Acid-core or rosin-core solder were not made for soldering (sweating) copper tubing.

Sandcloth or sandpaper should be used to clean the end of the tubing. Wire brushes are made for cleaning the inside of the fittings, but if you don't have one of these, wrap a piece of sandcloth or sandpaper around the eraser end of a pencil to clean the fitting. As I mentioned earlier, the tubing must be dry. If you can't shut the water off completely, if just a few drops seem to keep on coming

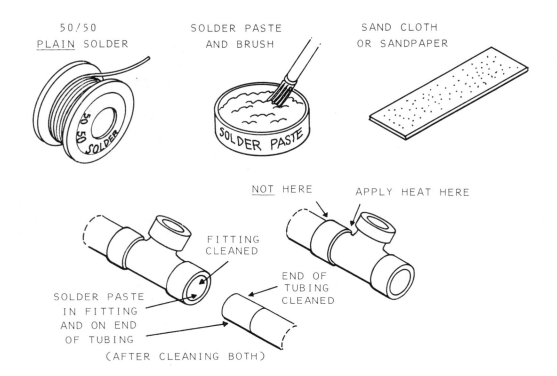

50/50
PLAIN SOLDER

SOLDER PASTE
AND BRUSH

SAND CLOTH
OR SANDPAPER

NOT HERE

APPLY HEAT HERE

FITTING
CLEANED

SOLDER PASTE
IN FITTING
AND ON END
OF TUBING

END OF
TUBING
CLEANED

(AFTER CLEANING BOTH)

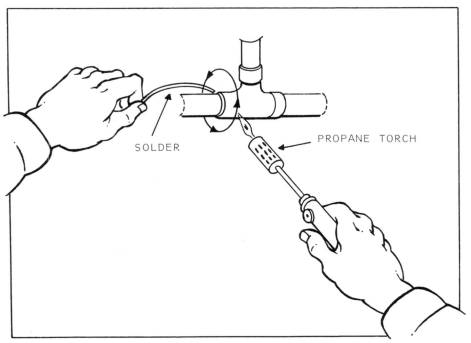

SOLDER

PROPANE TORCH

Fig. 153   Tools and materials needed to make sweat copper joints.

through, take a slice of bread, crumble it and stuff it back into the tubing—the eraser end of a pencil is a good tool for this—and as soon as you've stopped the dripping, solder the joint. When you've finished and turn the water back on, the bread will literally dissolve under pressure and can be washed out through a faucet or valve. Also, when soldering copper tubing, heat builds up in the piping and heat causes pressure. A valve or faucet must be open on the pipe being soldered to keep pressure from building up. If this is not done, the built-up pressure will blow small pinholes in the newly soldered joints, causing leaks when the water is turned on. If you follow these rules, you'll make perfect solder joints every time.

Chapter Thirty-Two

First Aid for a Damp Basement

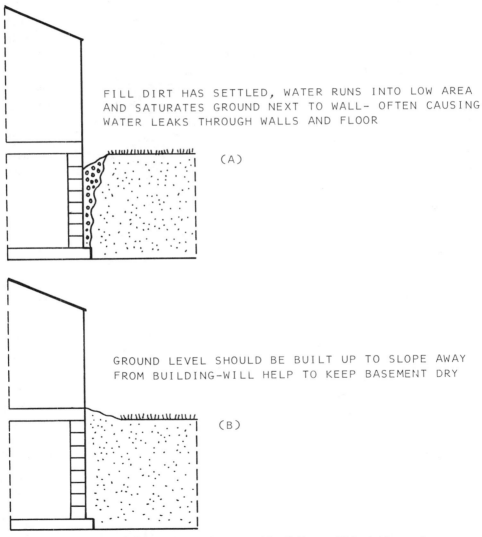

FILL DIRT HAS SETTLED, WATER RUNS INTO LOW AREA
AND SATURATES GROUND NEXT TO WALL- OFTEN CAUSING
WATER LEAKS THROUGH WALLS AND FLOOR

(A)

GROUND LEVEL SHOULD BE BUILT UP TO SLOPE AWAY
FROM BUILDING-WILL HELP TO KEEP BASEMENT DRY

(B)

Fig. 154   Proper ground level around building will help keep basement
dry.

One of the causes of damp or wet basements is settlement of the fill dirt around the foundation walls. The excavation for a basement must be at least 2 feet larger on each side than the outside of the finished wall. This is necessary to provide room for either the formwork for a poured-concrete wall or for room for the concrete-block layers to work if the walls are to be of block. After the foundation walls are built, fill dirt is then pushed into the space between the wall and the existing dirt bank. When the space has been filled the area is raked off level with the existing ground. As time goes by the fill dirt settles, leaving a low area next to the basement walls. The same settlement process can be seen wherever trenches were dug for sewer or water piping. Wherever dirt fill is used for backfilling excavations around buildings or for trenches, settlement will continue for years. In Fig. 154-A, we see how the fill dirt has settled next to the basement wall. Rain or melting snow will run to this low area, then seep down alongside the wall, and if the wall is not waterproof it will soak into and through the wall. In Fig. 154-B, the ground next to the wall has been filled in and built up slightly higher than the original ground level. Water will now run away from the house instead of soaking in around the basement wall.

Chapter Thirty-Three

Installing a Sump Pump

This project consists of 90 percent work and 10 percent know-how. If you will do the work, I'll furnish the know-how and show you how to go about it, step by step. The alternative is to hire a plumber to make the installation, but if you want to save a couple of hundred dollars, here's how to go about it. First of all, you'll need the following tools:

    4- or 5-pound hammer
    Cold chisel, 6 or 8 inches long, 1 inch in diameter
    2-inch hole saw (in mandrel)
    Small pointing trowel
    Shovel, spade or posthole digger
    Electric-drill motor and ½-inch drill bit
    6- or 8-pound sledgehammer (optional)

and the following materials:

    Sand-type concrete mix (can be purchased prepackaged)
    1 fiber-glass sump basin, with cover
    1 submersible sump pump
    1½ inch PVC or ABS Schedule 40 (DWV) plastic pipe
    MIP (male iron pipe) adapter, 1-1½ inch × 1¼ inch
    DWP-MIP (male iron pipe)
    1-1½ inch plastic vertical check valve, 1½ inch
    PVC or ABS cement

plus other assorted

    1½-inch DWV fittings as needed; the number and kind of fittings will depend on routing of drainage piping.

I recommend a submersible sump pump because with this type of pump there is no motor or lift rods above the floor. It makes for a much cleaner installation.

    The sump basin should be installed at what appears to be the wettest area of the floor and preferably near an outside wall. It should be located near a 110-volt electrical outlet (receptacle) if possible; this is not critical, however, as another outlet can always be added at this location if necessary. When you have selected the best location for the sump basin, turn the basin upside down and draw a line around the outside of the rim approximately 2 inches away from the rim. Place the basin (upside down) at least 6 inches away from the basement wall when marking the line; this should clear the wall footing. A few blows with a 6- or 8-pound sledgehammer in the center of the marked area will usually crack the concrete, after which a 4- or 5-pound hammer and the cold chisel can be used to break out the concrete to the marked line. After the concrete has been removed, the gravel fill and the sub-soil must be removed to a depth of approximately 32 inches, or the depth of the sump basin. When the concrete has been removed, exposing the gravel fill, save the gravel, as it can be used to backfill between the basin and the dirt wall after the basin is set. Placing the basin and marking the area to be cut and dug out is shown in Fig. 155. I should mention at this point that since you have a water problem the digging for the basin may have to be done underwater; this is *not* a problem, although it can be messy.

Fig. 155   Locating and marking hole to be cut in floor.

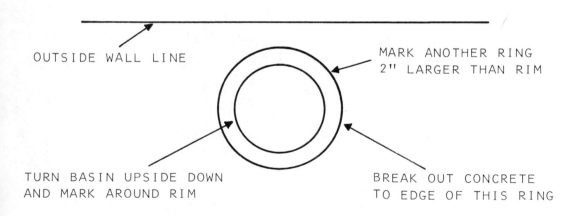

OUTSIDE WALL LINE

MARK ANOTHER RING
2" LARGER THAN RIM

TURN BASIN UPSIDE DOWN
AND MARK AROUND RIM

BREAK OUT CONCRETE
TO EDGE OF THIS RING

Try to keep the sides of the hole straight as you dig. I find a posthole digger the best tool for this purpose. Mark a 4- or 5-foot-long, 1 × 2-inch board to the depth of the basin and use this for a measuring stick to dig to the proper depth. As shown in Fig. 156, drill a series of ½-inch holes around the sides and at the bottom of the basin. The top row of holes should be approximately 6 inches below the rim of the basin. These holes will serve two purposes: they will allow the basin to be set initially—without the lower holes it is impossible to overcome the buoyancy of the basin when setting it—and if, after first setting it, you find that more digging is necessary, or if the hole is too deep and some filling is necessary, it will be easy to remove the basin.

Fig. 156   Setting the basin, backfilling and cementing floor.

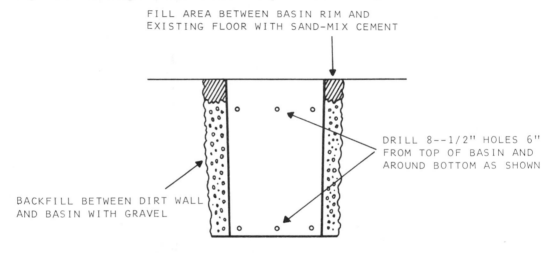

FILL AREA BETWEEN BASIN RIM AND
EXISTING FLOOR WITH SAND-MIX CEMENT

DRILL 8--1/2" HOLES 6"
FROM TOP OF BASIN AND
AROUND BOTTOM AS SHOWN

BACKFILL BETWEEN DIRT WALL
AND BASIN WITH GRAVEL

When the basin is properly set in place, the outside edge of the rim should be level with the surrounding floor. Gravel should be used to backfill the area between the basin and the excavated hole, as shown in Fig. 156, to within 4 inches of the existing floor. The sand-mix concrete will be used to cement the area between the rim and the concrete floor. The discharge piping from the pump could be connected to the sanitary building drainage piping if local regulations permit, but in my opinion the best way to dispose of the pumped-out water is to extend the piping through an outside wall and let the water flow onto a splash block and then be absorbed into the ground. There are several reasons for this, the most important being that if the piping is connected to the sanitary building drainage piping and the drainage piping should become clogged between the building and the sewer, the pump could force water up into the building and out of the fixture outlets, causing great damage. Also, since this type of installation is used only to remove ground water, this water is clear and odorless and can do no harm to the ground area which receives it. The diagram shown in Fig. 157 illustrates the suggested piping arrangement.

The purpose of the check valve in the discharge piping is to keep the water in the discharge piping from dropping back into the sump basin, where it would have to be picked up and pumped

Fig. 157   Preferred method of piping discharge from sump pump.

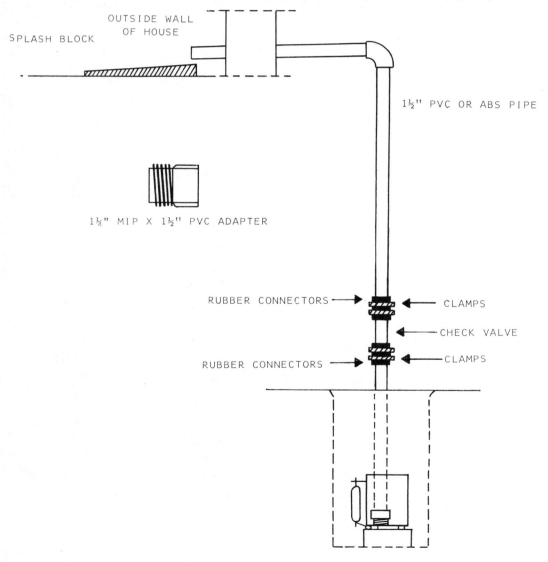

out again. The check valve is connected to the discharge piping by rubber connectors and clamps. This permits easy removal of the pump for cleaning of the inlet screen on the bottom of the pump.

To remove the pump loosen the lower clamp on the check valve and lift the check valve and upper piping approximately 1½ inches. The piping below the valve and the pump can then be lifted out of the sump basin. The pump should be removed and the inlet screen checked and cleaned at least once a year.

The 2-inch hole saw will be needed to cut through the wood header at the end of the floor joists where the discharge piping is extended outside the house. After the header is drilled out, a pilot hole can be drilled through this hole to the outside wall and the hole can be finished from the outside. The ground level outside must be lower than the point at which the discharge piping enters the outside. A splash block placed under the end of the discharge piping will break the force of the discharge and help prevent erosion. One-and-a-half-inch PVC or ABS DWV (Schedule 40) piping should be used for the discharge piping. Either type can be cut to length with a hacksaw or a cross-cut handsaw, and the joints between pipe and fittings are cemented together with PVC or ABS cement. A word of caution here: use only PVC cement with PVC pipe and ABS cement with ABS pipe. If one type of cement is used with the other type of pipe, the joints may not hold through the outside wall. And, with either cement, make certain the fittings are turned to the right angle or direction when you insert the pipe into the fitting. The cement will set almost instantly if the joint was made properly, and it will be impossible to turn the fitting on the pipe within 5 seconds. The proper way to cement PVC or ABS joints is to use the applicator in the can lid to spread the cement around the socket of the fitting, then apply cement to the end of the pipe and *insert the pipe into the fitting immediately.* The sump pump will have a 1¼-inch tapping for the discharge pipe; a 1½-inch PVC (or ABS) × 1¼-inch MIP (male iron pipe) adapter, shown in Fig. 157, will be needed to connect the discharge pipe to the pump. The check valve should be placed about 12 inches above the basin cover.

A plumbing wholesale house is the best place to buy the pump, basin, cover, piping and fittings. The supply house will also have most, if not all, of the tools and other materials you need for this job. Other sources for these materials are hardware or building-supply stores.

Chapter Thirty-Four

Maintenance of a Gas-Fired Forced-Air Furnace

As an aid to maintenance procedures, let's quickly run through the sequence of operation of a typical gas-fired blower furnace. The pilot burner is on, the temperature at the thermostat drops below the set point of the thermostat. The drop in temperature closes a switch in the thermostat, completing an electrical circuit to the main gas-burner valve on the furnace and permitting gas to flow through the valve to the burner. As the gas leaves the burner it is ignited by the pilot flame and heat starts to build up in the heat exchanger in the furnace. When the heat reaches the high setting of the fan switch, the fan switch is turned on, completing an electrical circuit to the blower motor. The blower is then turned on and heated air is circulated through the house. The burner can make heat faster than the air circulated by the blower can remove it from the heat exchanger, so when the temperature in the heat exchanger reaches the setting of the limit switch the limit switch opens, breaking an electrical circuit to the main gas burner and shutting off the gas to the main burner. The blower will continue to operate until the temperature in the heat exchanger reaches the low setting of the fan switch, at which point the fan switch opens and the blower is shut off. This sequence of operation will continue until the desired room temperature at the thermostat is reached.

Clean or Replace Air Filters Regularly
When air filters become dirty, airflow through the filters is drastically reduced. The furnace blower is rated to move x number of cubic feet per minute at x number of motor rpm. As I explained in the sequence of operations, the blower will remain in operation until the heat has been removed (thus transferred into the home) from the heat exchanger. Dirty filters will cause the blower to run longer than necessary, wasting electricity—and dollars!

Clean the Vanes on the Furnace Blower

Squirrel-cage blowers are used in furnaces because they are quieter and more efficient than fan types. The vanes are the slots which move air through a squirrel-cage blower. Dust and dirt particles can build up on these vanes and reduce the amount of air the blower will move. Once a year, before the start of the heating season, the blower should be inspected and the bearings (unless they are the sealed type) on the motor and the blower shaft should be oiled. Any dust or dirt on the blower vanes should be removed. An old toothbrush can be used to loosen the dust and dirt, which can then be removed from the blower and blower compartment with a vacuum hose. The main electrical switch controlling the furnace should be turned off before working on the blower, cleaning or changing the filters or the burners.

Clean the Burners and the Burner Compartment

Rust and scale are normal by-products of burner operation. The burners should be removed and brushed with a wire brush, and scale and rust should be removed from the blower compartment. This should be done every year before the start of the heating season.

How to Check for a Burned-Out or Rusted-Out Heat Exchanger

A burned-out or rusted-out heat exchanger will permit *carbon monoxide*, a by-product of burner operation, to enter the home. Carbon monoxide is *odorless, colorless, poisonous* and *deadly*! Every gas furnace, regardless of its age, should be checked regularly to ensure that the heat exchanger is in good condition. As I mentioned earlier, in a normal heating cycle the main burner will come on and when heat builds up in the heat exchanger the fan switch will turn the blower on. One good check on the condition of the heat exchanger is to watch the burner flame when it *first comes on*. The flame will waver at first, then become steady. Watch the flame carefully until the blower comes on. If the flame starts to dance or show movement or change of color, it is an indication that air from the blower is affecting the flame. If air from the blower can escape from the heat exchanger through a burned-out or rusted-out hole, then carbon monoxide can enter the heat exchanger (and the home) through the same hole. If you find that the flame is disturbed when the blower comes on, find the gas valve controlling the gas to the furnace and turn off this valve. Call your local gas utility or a qualified serviceman to have them inspect the furnace. *Remember, carbon monoxide is deadly!*

Should You Leave the Pilot Light on in a Furnace During the Summer?

In my opinion, yes. The heat from the pilot light, small though it is, will keep condensation from forming on, or in, the heat exchanger. If the heat exchanger is dry, little or no rust will form.

Check the Vent Pipe Between the Furnace and the Chimney

Inspect the vent pipe before the start of the heating season. If the pipe is rusty or corroded, it should be replaced. Feel along the bottom of the pipe; acids formed by combustion by-products and moisture will often eat away the bottom of the vent pipe without affecting the appearance of the rest of the pipe. If a central air-conditioning unit is incorporated in the furnace, the blower will be operating throughout the year and the air filters should be cleaned or replaced during the summer in order to have an efficient unit.

Chapter Thirty-Five

Preventing Damage to Outside Hose Faucets

The hose faucet (sill cock) shown in Fig. 158 is called a freezeless or antifreeze sill cock. It will be an antifreeze sill cock if it is installed correctly and if the hose is removed from the sill cock before the outside temperature dips below freezing. This type of faucet was designed to be used with the shutoff point inside the building walls in a warm area. The faucet is made with a long rod attached to the handle; the faucet washer and shutoff point are at the end of the rod. When the sill cock is shut off any water remain-

Fig. 158   A freezeless type of sill cock.

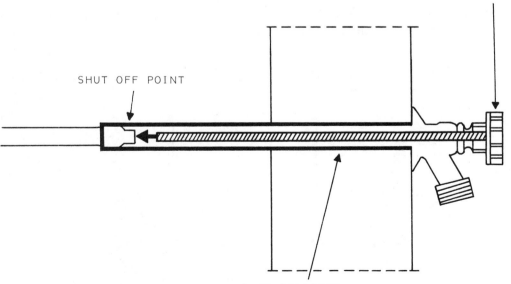

IF SILLCOCK IS A FREEZELESS TYPE, WHEN THE HANDLE IS CLOSED
WATER WILL DRAIN FROM THE SILLCOCK IF THE HOSE IS REMOVED

SHUT OFF POINT

WATER WILL BE HELD IN SILLCOCK
IF HOSE IS LEFT ATTACHED

ing in the body of the sill cock will drain out of the spout *if the spout is open*. If a hose is left attached, the water cannot drain out and will freeze when subjected to freezing temperatures. If the stem and handle are at a 90-degree angle to the wall, as shown in Fig. 158, the sill cock is probably an antifreeze type. If, as in Fig. 159, the handle and the stem are at a 45-degree angle (±) to the wall, the sill cock is not an antifreeze type and it must be drained to prevent damage in below-freezing temperature.

When this type of sill cock is used a shutoff valve, called a stop-and-waste valve, should be installed in the piping going to the sill cock. To protect the sill cock from freezing, turn the stop-and-waste valve off, open the sill cock and unscrew and remove the small cap on the side of the stop-and-waste valve. When the cap is removed a small hole will be visible, air will enter through this hole and break the vacuum in the piping to the sill cock and water in the piping or sill cock will drain out if the hose has been removed. Occasionally the small rubber washer on the inside of the cap will stick to the valve when the cap is removed. If this should happen the washer will pull off easily and must be re-placed back into the cap. After the water has drained from the sill cock, replace the small cap on the valve but do not tighten it. Leave it loose until the sill cock is needed in the spring. Leave the stop-and-waste valve shut off until danger of freezing is past.

New washers can be installed on either type of sill cock by removing the packing-gland nut under the handle, turning the handle counterclockwise to remove the stem and replacing the washer on the end of the stem.

Fig. 159   A standard type of sill cock with stop-and-waste valve inside the house.

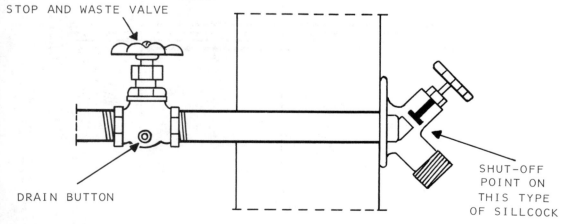

STOP AND WASTE VALVE

DRAIN BUTTON

SHUT-OFF POINT ON THIS TYPE OF SILLCOCK

3
HOME
IMPROVEMENTS

Chapter Thirty-Six

Tub and Shower Enclosures

The perfect finishing touch to create an attractive bathroom is to install glass doors on the tub and shower or a separate shower. Shower curtains after a time tend to become moldy and dingy around the bottom. Unless they are fully closed they often permit water leakage, which if allowed to continue can eventually damage the floor. The best of shower curtains has a fairly short life

Courtesy Southeastern Aluminum Prods, Inc.

Fig. 160   Sliding glass doors for tub and shower.

Courtesy Southeastern Aluminum Prods, Inc.

Fig. 161   A hinged glass door for stall shower.

expectancy, and for about the cost of three good shower curtains a homeowner, doing the work himself, can install sliding glass doors on the tub or a hinged glass door on a shower. The installation time is about 2 hours for either one and the installation is not difficult; full instructions are packaged with each unit. Holes to mount the side rails will have to be drilled through the walls; in almost all cases these holes must be drilled through ceramic tile. They can be drilled very easily using the type of masonry drill shown in Fig. 57.

Glass doors are made of either tempered glass or polystyrene plastic set in polished-aluminum frames. They will retain their bright shiny appearance for many years with only a minimum of care. Soap film will wipe off easily with a damp sponge. A tub and shower door is shown in Fig. 160; a hinged door for a separate shower is shown in Fig. 161.

Chapter Thirty-Seven

Ceiling Fans

Ceiling (paddle) fans are energy savers, so much so that in many cases they can repay their cost in the first three months of use. In the summer ceiling fans make air conditioners more efficient by circulating cool air. During cold weather the fan circulates warm air which would otherwise be trapped near the ceiling. In marginal air-conditioning weather the fan alone can often be used to provide relief from the heat, and the fan, which has a very small motor, will operate at a fraction of the cost of an air-conditioning unit.

Ceiling fans are made for use either with concealed wiring into a ceiling electrical box or with swag kits, in which the wiring is extended through a decorative chain hung from the ceiling and then dropped down the face of a wall to an electrical receptacle.

Ceiling fans are either made with a pull chain connected to a multispeed switch mounted in the fan housing to control fan

Courtesy NuTone Division

Fig. 162   An energy-saving ceiling fan.

speed or are designed to use a wall-mounted multispeed switch to control fan speed.

If a light kit is to be installed on a ceiling fan, a separate 110-volt power supply should be installed in the ceiling electrical box to operate the lights. Using a typical wall-mounted multispeed switch to operate the fan the sequence of wiring is: from a single-pole wall switch to the multispeed switch, then to the ceiling box and fan. This means that the speed-control switch is in series with the fan and if lights were also connected to this circuit, when the fan's speed is reduced the voltage to the lights would also be reduced, dimming the lights. Also, the lights could be turned on only when the fan is in operation. For this reason two "hot" wires are required when a fan uses a wall-mounted speed controller. Two 110-volt cables are shown in the ceiling box in Fig. 81. A typical ceiling fan is shown in Fig. 162. Three- and four-blade models are available with blade sweeps of from 36 to 54 inches.

Installation is simple and easy but does require a solid mounting support, similar to the one shown in Fig. 81, into which the mounting hook must be inserted. When swag-type fans are used the mounting hook can be screwed into a ceiling joist or other solid-wood framing member.

A substantial reduction in both heating and cooling costs can be made by installing and using ceiling fans.

Complete installation instructions are furnished with the units.

Chapter Thirty-Eight

This Energy-Saving Device Saves Money, Too

There is a very simple and effective way to save energy and money in electric water heater operation. Install an automatic timer to control the operation of the heating elements in the heater. This device, shown in Fig. 163, has a 24-hour clock and can be set to turn on to provide hot water when needed and off when not needed. It can be set to provide hot water in the morning, turn itself off when not needed during the daytime hours, turn itself on in time to provide hot water as needed during the evening and then shut off during the night hours. By means of operators mounted on the clockface, the homeowner can set the timer to operate at any desired times.

Tests using the timer were made in Illinois and in Florida, two areas where the incoming-water temperature and the recovery time vary greatly. Recovery time means the time it takes to raise the temperature of x number of gallons of water 100 degrees F.; or, to put it another way, the recovery rate of electric water heaters can be figured on the basis of 4.1 gallons per hour per kilowatt (1,000 watts) at 100 degrees F. temperature rise. If the incoming-water temperature is 60 degrees and the thermostats on the heater are set at 140 degrees, the heater must raise the incoming water temperature 80 degrees. Most of the water heaters in use have two 4,500-watt elements, only one of which operates at one time. Using the formula given above, the heater will recover, or heat, about 19.5 gallons per hour.

Incoming water is much warmer in Florida than in Illinois, so for tests to be accurate showing an overall average savings, both areas should be taken into account.

In Illinois, when the Intermatic timer was installed, the savings in kilowatt-hours of electricity *used to heat water* varied from 30.2 to 38.7 percent, for an overall average of 35.1.

Fig. 163   The energy-saving Intermatic timer.

In Florida the test results varied from a high of 40.6 to a low of 27 percent, an overall average savings in kilowatt-hours of electricity *used to heat water* of 33 percent. These results do not show the percentage of savings on the total electric bill, they reflect only the savings effected in heating water.

From these test results, assuming a national average cost of .048 per kilowatt-hour for electricity and a $50 installed cost for the Intermatic timer, in Illinois the timer would pay for itself in savings in 5 months; in Florida it would pay for itself in 6.6 months.

The timer can be switched manually to either on or off position without changing the operators on the clockface. This feature enables the homeowner to have hot water during periods of heavy usage, guests in the home, holidays, large amounts of laundry, etc. The manual switching of the timer to on or off overrides the operational mode of the timer at the time the manual switch is made. At the next preset period the timer will take over and go into automatic sequencing again.

Example: Your timer is set to go to on position at 6 A.M. and to off position at 8 A.M. This setting would furnish ample hot water for morning baths, etc. The timer would remain in off position until 3 P.M., then go to on position and remain on until 7 P.M., at that time automatically switching to off. Knowing the hot-water usage will be heavy today, you manually switch the timer to on position at 9 A.M. The heater will automatically continue to heat water as needed until 7 P.M. At 7 P.M. (the preset time) the timer will take over, switching to the off position, and continue in automatic operation.

In certain areas of the country where the water is "hard," scale is precipitated out of the water when the water is being heated. This scale will settle to the bottom of the heater and if not removed will eventually cover the lower element, causing it to burn out and fail. It follows then that operating the heater only when needed can add to the lifetime of the elements.

INSTALLING THE TIMER

I recommend that unless you are very knowledgeable about electricity you have an electrician install the timer. You will be working with somewhere between 217 and 240 volts and a mistake could be fatal. If you feel confident that you can install the timer, take this precaution: turn off *all* electrical power to the building. After you've turned it off, check lights, receptacles, the electric stove, etc. to make certain that the power is *off*. If the breakers or fuses controlling the power to the heater are marked, turn them off or, in the case of fuses, remove them as an added precaution. I have known breakers or fuses to be mislabeled, so even though the breakers or fuses are marked, leave the main switch, breaker or fuse *off* until the job is completed.

With all the electric power off, remove the plate at or near the top of the heater where the wiring enters the heater. Disconnect the incoming-power wires from the heater wires; the bare or ground wire should be connected to a green or hex screw. Disconnect the ground wire also and remove the cable containing all wiring from the heater.

Mount the timer on a wall close to the heater, at a point where the power cable will reach the timer. Connect the two power wires (two reds, two blacks or a red and black), one wire to line 1 terminal, the other to line 2 terminal. Do not connect the bare (ground) wire at this time.

It may be necessary to buy a short length of wire to connect from the timer to the heater; this wiring should be no smaller than number 10, and local codes may require that it be run in armored cable or conduit. The wiring should be the 2 wire with ground type. One wire should be connected to the load 1 terminal, the other to the load 2 terminal. The two bare ground wires should be twisted together with a third piece of bare ground wire, as shown in Fig. 164, and connected to the ground (green or hex) screw in

Fig. 164   Electrical wiring to timer.

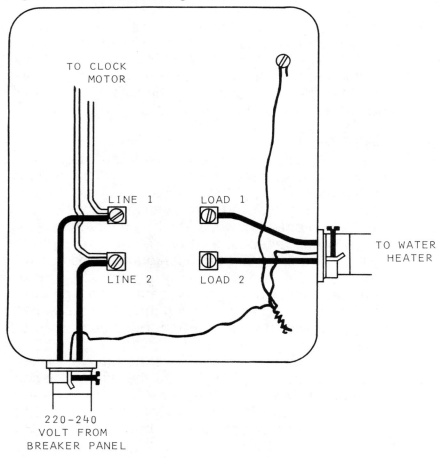

the timer box. Connect the other ends of the red, black or red and black, wires to the wiring in the heater which was disconnected earlier. Colors of the wires are not important here, *with this exception*: green wires are always ground wires and must be connected to the ground connection of the appliance. Otherwise the red and black wires (or a combination of these) should be connected as they were originally. The reason for this is that each incoming-power wire carries 120 volts, and when each wire is connected to a wire at the heater the combination of these two wires totals 240 volts. Wire nuts are commonly used to make connections at appliances. When they are used about ½ inch of the insulation should be removed from the end of each wire, then the two (or more) wires should be twisted tightly together, using pliers. The wire nut can then be twisted onto the wires, making a tight connection, with no bare wire visible.

When wiring is connected to terminals, the screws must be tightened as tight as possible for a good connection.

After completing all connections at the timer and the water heater, inspect all wiring and connections: be certain all wires are connected correctly, then replace plates, covers, etc. After all openings are secured, turn the breakers controlling the electric power to on position (or replace fuses).

Directions for setting the timer and placing the actuators on the timer face may vary according to the timer purchased.

Chapter Thirty-Nine

Solar Water-Heating Systems

As fuel costs continue to rise, the installation of solar water-heating systems becomes more practical and desirable with every passing day. The energy we buy is very expensive; the energy from the sun is free. A question which always comes up at this point is, "What happens on a cloudy day?" Even on cloudy days some energy is available and is collected by solar panels. Standby electrical heater elements will automatically take over to ensure that a supply of hot water is available when needed. Another frequently asked question is, "Does the solar heating system work when the outside temperature is below freezing?"

The answer is emphatically *yes*. Collector panels are designed for freeze or nonfreeze areas. The sun's rays penetrate the glass cover and are absorbed by the black-surfaced absorber plate. This in turn heats the transfer liquid, which is an antifreeze mixture of propylene glycol and water. The hot liquid is then circulated to the copper-coil heat exchanger in the water-heater tank, where the heat from the hot liquid is transferred to the water stored in the heater.

What about the cost of installation and operation of a solar water-heating system? If it is installed correctly and maintained properly the system should pay for itself in a very few years. This is not a do-it-yourself project. The installation of a solar water-heating system should be done by an expert in this field if the owner of the system is to get the full benefits of it. The placement and adjustment of the solar panels is critical to the successful operation of the system.

The number of panels needed will vary with the individual system. The system's manufacturer can recommend a qualified master plumber to select the materials and make the installation for you.

The installation costs will depend on the number of solar panels needed, the amount of on-the-job piping and the job site conditions. Federal tax credits can help offset the initial installation costs. Here's how the tax credit works:

At the time I am writing this a federal tax credit of 30 percent on the first $2000 and 20 percent on the next $8000 or any part thereof is allowed. Assuming a cost of $2400 for the installation, a tax credit of $680 is allowed (30 percent of $2000 = $600 + 20 percent of $400 = $80—for a total tax credit of $680).

> Total cost _____$2400
> minus tax credit_____ 680
> cost after credit_____$1720

If this installation is in lieu of a standard-type-heater installation, you would be saving an additional $300 (±), or the installed cost of the heater. If you wish to install a conventional heater to use as an additional storage tank in a two-tank system, the above additional saving would not apply.

Let's get to the mechanics of a solar water-heating system. Fig. 165 shows the installation of a system using two solar collector panels connected to a water heater which has supplementary electric heating elements. These elements are designed to go into operation as backup units, to ensure the supply of hot water in the event of a system failure or during periods of heavy usage. In Fig. 166 we see the same system with the addition of an auxiliary storage tank to provide a larger volume of available hot water.

The cutaway drawing in Fig. 167 shows the construction of a typical solar panel. Here you can see how heat is absorbed by the black-surfaced copper absorber plate; this heat is transferred to the fluid which is circulating in the panel's tubing. You will notice that with this type of system the water in the tank is not heated directly by the panels. It is heated *indirectly* by the hot heat-transfer liquid circulated from the panels to the heat-exchanger coils in the heater and then back to the panels by a pump mounted on top of the heater. An automatic air vent is installed to purge air from the heat-transfer liquid's piping, and a diaphragm-type expansion tank is installed to allow for the expansion of the heat-transfer liquid as it is heated in the solar panels. Piping for the solar panel is shown in Fig. 168. Controls are provided to prevent overheating of the water stored in the heater tank, either from solar heating or from the electric elements in the heater. The prices used in the above example are for explanatory purposes only and do not represent actual installation costs.

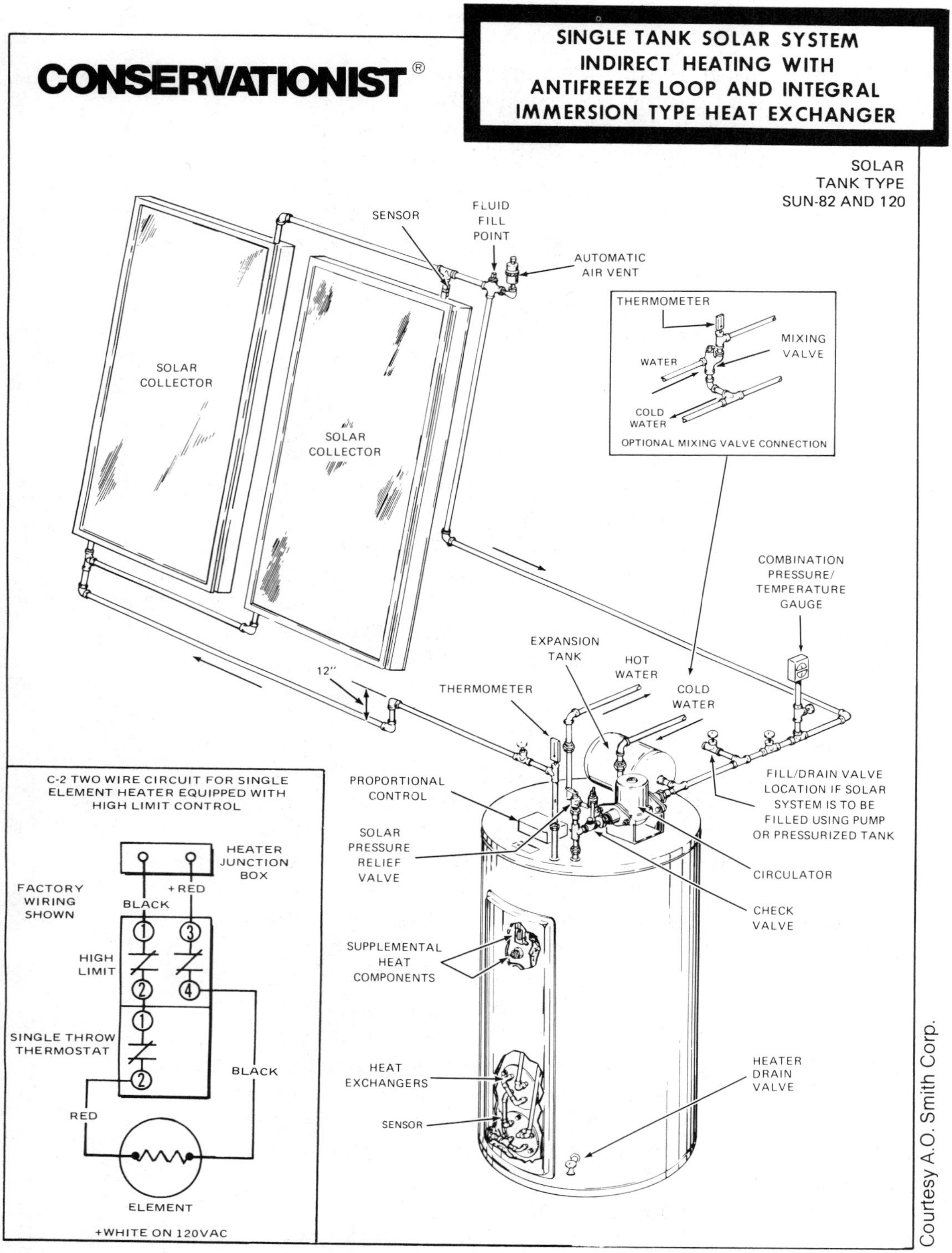

CONSERVATIONIST ®

SINGLE TANK SOLAR SYSTEM
INDIRECT HEATING WITH
ANTIFREEZE LOOP AND INTEGRAL
IMMERSION TYPE HEAT EXCHANGER

SOLAR
TANK TYPE
SUN-82 AND 120

SENSOR

FLUID
FILL
POINT

AUTOMATIC
AIR VENT

SOLAR
COLLECTOR

SOLAR
COLLECTOR

THERMOMETER

MIXING
VALVE

WATER

COLD
WATER

OPTIONAL MIXING VALVE CONNECTION

COMBINATION
PRESSURE/
TEMPERATURE
GAUGE

EXPANSION
TANK

HOT
WATER

COLD
WATER

12"

THERMOMETER

FILL/DRAIN VALVE
LOCATION IF SOLAR
SYSTEM IS TO BE
FILLED USING PUMP
OR PRESSURIZED TANK

C-2 TWO WIRE CIRCUIT FOR SINGLE
ELEMENT HEATER EQUIPPED WITH
HIGH LIMIT CONTROL

PROPORTIONAL
CONTROL

SOLAR
PRESSURE
RELIEF
VALVE

CIRCULATOR

HEATER
JUNCTION
BOX

CHECK
VALVE

FACTORY
WIRING
SHOWN

BLACK

+ RED

HIGH
LIMIT

SUPPLEMENTAL
HEAT
COMPONENTS

SINGLE THROW
THERMOSTAT

BLACK

RED

HEAT
EXCHANGERS

SENSOR

HEATER
DRAIN
VALVE

ELEMENT

+WHITE ON 120VAC

Courtesy A.O. Smith Corp.

Fig. 165   A single-tank solar water-heating system.

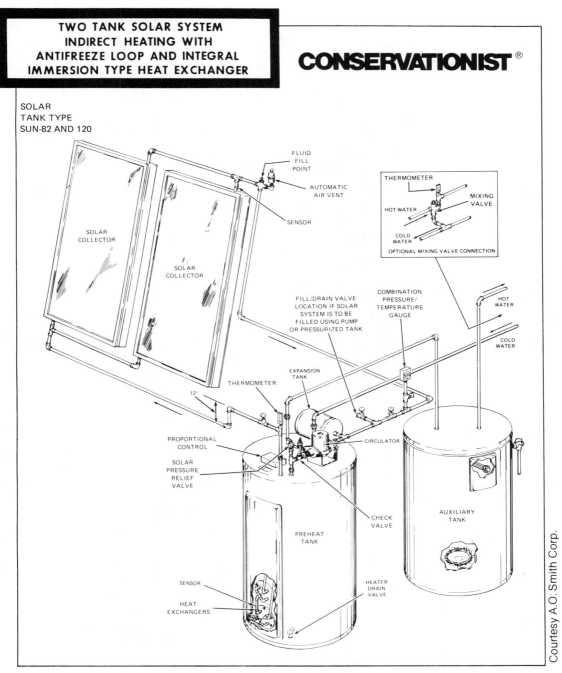

TWO TANK SOLAR SYSTEM INDIRECT HEATING WITH ANTIFREEZE LOOP AND INTEGRAL IMMERSION TYPE HEAT EXCHANGER

CONSERVATIONIST ®

SOLAR
TANK TYPE
SUN-82 AND 120

FLUID
FILL
POINT

AUTOMATIC
AIR VENT

SENSOR

SOLAR
COLLECTOR

SOLAR
COLLECTOR

THERMOMETER

MIXING
VALVE

HOT WATER

COLD
WATER

OPTIONAL MIXING VALVE CONNECTION

FILL/DRAIN VALVE
LOCATION IF SOLAR
SYSTEM IS TO BE
FILLED USING PUMP
OR PRESSURIZED TANK

COMBINATION
PRESSURE/
TEMPERATURE
GAUGE

HOT
WATER

COLD
WATER

EXPANSION
TANK

THERMOMETER

12"

PROPORTIONAL
CONTROL

SOLAR
PRESSURE
RELIEF
VALVE

CIRCULATOR

CHECK
VALVE

AUXILIARY
TANK

PREHEAT
TANK

HEATER
DRAIN
VALVE

SENSOR

HEAT
EXCHANGERS

Courtesy A.O. Smith Corp.

Fig. 166   A two-tank solar water-heating system.

227

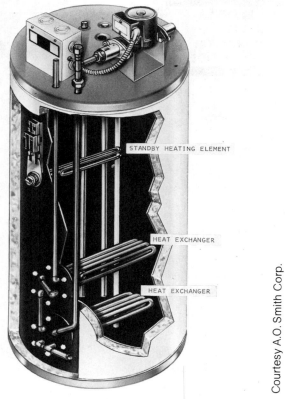

STANDBY HEATING ELEMENT

HEAT EXCHANGER

HEAT EXCHANGER

Fig. 167   Cutaway drawing of A. O. Smith storage tank.

Fig. 168   Piping to and from solar collectors on roof.

TYPICAL PLUMBING ARRANGEMENT

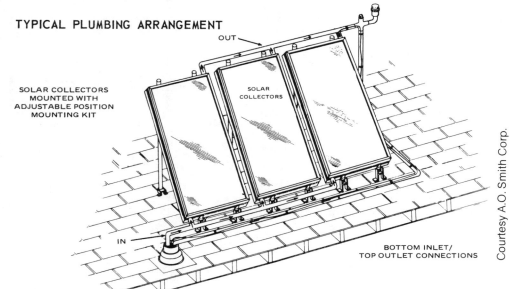

OUT

SOLAR COLLECTORS
MOUNTED WITH
ADJUSTABLE POSITION
MOUNTING KIT

SOLAR
COLLECTORS

IN

BOTTOM INLET/
TOP OUTLET CONNECTIONS

Chapter Forty

Tired of Your Old Floors?

Do-it-yourself tiles make floor replacement easy. Most homeowners consider the replacement of kitchen flooring a major expense, and indeed at one time it was. Not only did you have to pay the cost of the new floor, but there was substantial extra expense for the labor involved in having it installed: preparing the old floor, spreading adhesive and cutting and fitting the new floor. Today the job is a breeze thanks to the new do-it-yourself floor tiles that come with their own adhesive preapplied to the back. Doing the work yourself, it's possible to cover a 10 × 15-foot kitchen in less than 4 hours. And because you do the work yourself, you save the cost a professional installer would charge.

The key difference between these do-it-yourself tiles and conventional floor tiles is there's no adhesive to spread. It's already on the back of the tile. You simply peel off a protective paper backing to expose the adhesive, as shown in Fig. 169, place the tile in position on the floor and press down firmly. Three of the most popular kinds of self-adhering tiles are vinyl-asbestos, no-wax and vinyl tiles that contain no asbestos filler. These tiles can be cut and applied as shown in Fig. 170 and Fig. 171.

A STAPLE-DOWN VINYL FLOOR
A lot of people, certainly all true-blue do-it-yourselfers, know how to install vinyl flooring. You slather adhesive all over the surface you're going to cover, then you lay the new vinyl floor down over the top. Right? Well, that's one way to do it. But there's a far easier way to install vinyl flooring: staple it, as shown in Fig. 172. If you think you didn't read the last sentence right, you did. Because now there's a new type of cushioned vinyl floor from Armstrong that you simply staple down. And you only have to staple it at the sides of the room.

If you're applying this new floor over concrete, you apply a narrow band of cement at the room edges instead of stapling. In either case you no longer need to be bothered with spreading

Fig. 169   Peeling the backing from self-stick tile.

Fig. 170   To apply the tile, lay it on the floor and press down firmly.

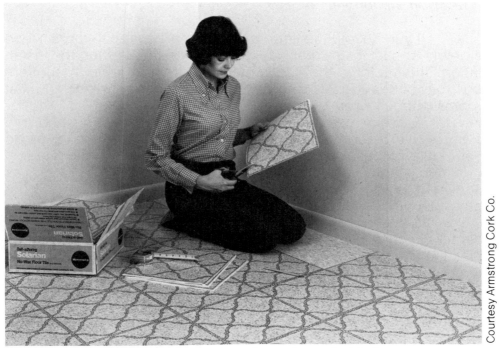

Courtesy Armstrong Cork Co.

Fig. 171   The tile can be cut with ordinary household shears.

adhesive. This new vinyl floor is made differently from all previous vinyl floors, so it performs differently. And it has other advantages, too, besides the unique way it's installed. It's an unusually flexible floor; you can fold it like a blanket and tuck it in the back seat of your car to take home from the store. This flexibility makes it easier for the amateur to work with, to maneuver, trim and fit. It's not at all stiff or "boardy" and it's elastic. If the floor is slightly undercut it can be stretched. It's the first vinyl floor that forgives a do-it-yourselfer's mistakes—up to a point. Most remarkable of all, this floor has a built-in "memory" (in the technical sense). When the floor is rolled face side out at the factory for shipment, the outer circumference of the roll is stretched. After installation in your home, the floor gently contracts as it seeks to return to its original dimensions before it was rolled up. This causes any slack or wrinkles you might have left in the material during the installation to be taken up gradually by the floor's memory action. The memory also protects the floor from expansion and contraction of the sub-floor caused by seasonal changes in temperature and humidity. Other vinyl floors under the same conditions might crack or buckle or separate at the seams. This new one follows the movements of the subfloor to stay tight and smooth season after season.

If you do fold the floor to take home, be sure to reroll it face side out and let it stay that way overnight before starting the installation. This restores the floor's memory. It is available in continuous rolls of 6- and 12-foot widths; the 12-foot width will cover most rooms without a seam.

To install, unroll the floor in the room (Fig. 173) and trim away any excess material, using a metal straightedge to guide your knife blade, as shown in Fig. 174. Then, simply take a staple gun and staple the floor at 3-inch intervals along the walls, as shown in Fig. 172. You can hide the staples with quarter round or other decorative molding.

If you're installing the floor over concrete, the proper cement in an applicator bottle is available at the dealer's. You can install this vinyl floor over almost any kind of surface anywhere in the house, from basement to attic. It can be installed over floors in poor condition—floors that otherwise would require sanding and patching or leveling. However, if there is major damage to the subflooring you will have to repair it before installing the vinyl. The vinyl floor has a tough wear layer of clear vinyl that resists stains and is easy to clean.

Fig. 172   Stapling down a floor eliminates the mess of gluing.

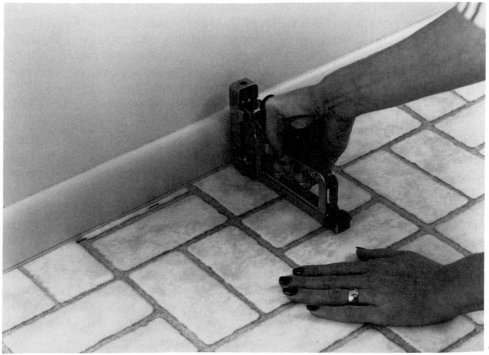

Courtesy Armstrong Cork Co.

Fig. 173   Unrolling the new vinyl floor.

Fig. 174   Care in cutting ensures a tight fit at wall.

Chapter Forty-One

An Energy-Saving Fireplace

There was a time not too long ago when a fireplace had to be built at the time the house was under construction. Modern building methods, new materials and expert engineering by manufacturers now make it possible to add safe, energy-efficient fireplaces to existing homes or home additions. Also, fireplaces were considered wasteful of the energy released in the burning process and in allowing heat from the home to escape up the chimney during periods when the fireplace was not in use. These problems can be solved by the addition of heat exchangers built into the fireplace. The heat exchanger takes cold air from the room, heats it and returns it to the room. The point of return can be close to the fireplace or ducted to a remote area of the room. The circulation of air, which will occur naturally by convection, can be made very efficient by the use of fans available as an add-on accessory kit. Also, two things can be done to prevent the loss of heated air in the home through the fireplace chimney. One is to keep the damper shut when the fireplace is not in use, two is to install a glass screen in the fireplace.

The fireplace is shown installed and in use in Fig. 175. All necessary chimney piping and accessories including the top housings are available from the supplier.

How efficient is the energy-efficient fireplace shown in Fig. 176? Heat output refers to the total amount of heat recovered from a specific weight of fuel being burned in the fireplace. It's actually a combination of two different kinds of heat: radiant heat and convective heat. Radiant heat is the heat projected directly into the home from the firebox. Convective heat is the heat that circulates into the home from the movement of air around the firebox. For example, a fire that consumes 13 pounds of wood per hour (a normal fire) in the Energy Saving TM heat-circulating fireplace will produce 18,000 BTUs per hour of radiant heat. At the same time, the fire will also produce 20,500 BTUs per hour of convective heat—a total heat output of 38,500 BTUs per hour. Increasing the amount of wood burned will of course increase the heat output.

A measurable mark of quality of a product, especially one which is readily adaptable to a do-it-yourself project, is the type and clearness of the assembly instructions. The illustrations in Figs. 177 and 178 are not included in this book to show you how to install the fireplace and its assorted accessories. They are included to show you how completely the assembly and installation instructions cover the job. A booklet of instructions is furnished with the fireplace, complete with step-by-step instructions for the assembly and installation of the energy-efficient fireplace.

Fig. 175　A fireplace can be a beautiful and valuable addition to a home.

Courtesy The Majestic Company

Fig. 176   A Majestic energy-saving fireplace
ready to set in place and connect.

Fig. 177   Complete rough-in measurements are furnished with the unit.

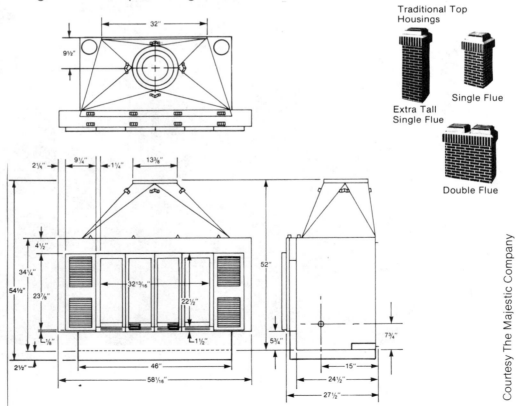

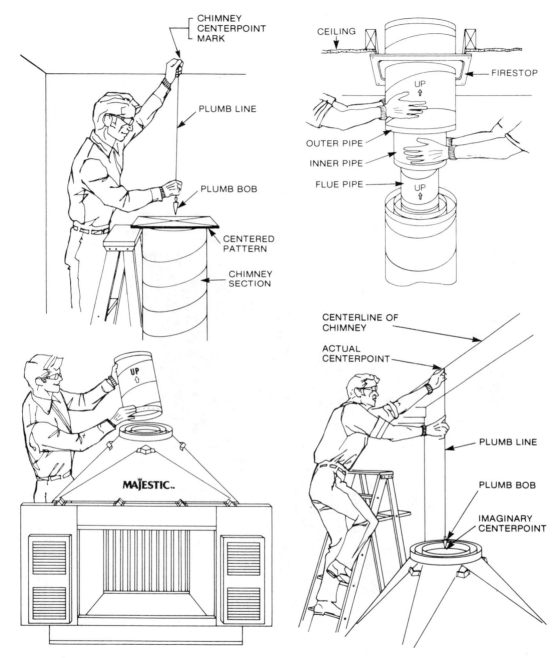

Fig. 178   Complete installation instructions are furnished by the manufacturer.

Courtesy The Majestic Company

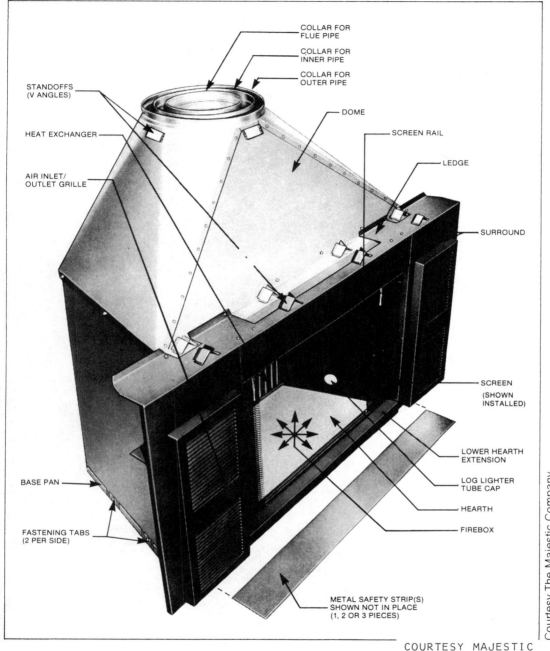

COLLAR FOR
FLUE PIPE

COLLAR FOR
INNER PIPE

COLLAR FOR
OUTER PIPE

STANDOFFS
(V ANGLES)

DOME

HEAT EXCHANGER

SCREEN RAIL

AIR INLET/
OUTLET GRILLE

LEDGE

SURROUND

SCREEN
(SHOWN
INSTALLED)

BASE PAN

LOWER HEARTH
EXTENSION

LOG LIGHTER
TUBE CAP

FASTENING TABS
(2 PER SIDE)

HEARTH

FIREBOX

METAL SAFETY STRIP(S)
SHOWN NOT IN PLACE
(1, 2 OR 3 PIECES)

Fig. 179   Parts identification of the energy-saving heat-circulating fireplace.

Chapter Forty-Two

A Safe, Automatic Garage-Door Opener

Courtesy NuTone Division

Fig. 180   The unit is packaged for easy pickup and transportation.

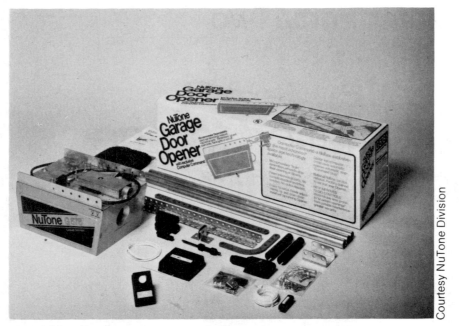

Fig. 181　Door-opener parts laid out for easy assembly.

Fig. 182　The NuTone automatic door opener assembled.

The safety and comfort feature of being able to open the garage door without getting out of the car makes the installation of an automatic opener well worth considering. The computer-controlled chain-drive garage-door operator is packaged for easy transportation and assembly, as shown in Fig. 180. The receiver and one transmitter are included. The microcomputer solid state circuitry provides an exclusive memory brain to signal commands for up and down limits of the door, automatic reverse and timed lights. The automatic reverse is what I consider the most important feature of this door. If the door meets resistance while going down (a small child or animal or even the trunk of a car), it automatically goes into reverse. The feature could save a life or prevent serious injury or property damage. Complete and easy to follow instructions make it possible to make short work of assembling the parts in Fig. 181 into the finished installed opener as shown in Fig. 182. The unit is made by NuTone.

Chapter Forty-Three

A New Kitchen at an Affordable Cost

Many kitchens in older homes are larger and could be much easier to work in than kitchens in new homes. If your kitchen were remodeled, with new cabinets, a new sink and counter top and a new floor, you could have a new kitchen without buying a new home. The ideas incorporated in new cabinets can make a kitchen a joy to work in. Visible storage in tall cabinets, shown in Fig. 183, brings everything in the cabinets within easy reach. Multistorage pantries (Fig. 184) are made with adjustable shelves; with this pantry in your kitchen it's like having your own private grocery store.

The pull-out tabletop in Fig. 185 can be used whenever additional work surface is needed.

The double swing-outs in Fig. 186 bring pans and groceries out into the kitchen, right to your fingertips.

And finally, the glide-out wastebasket (Fig. 187) is easy to get to yet out of the way when not needed. New cabinets which are designed not only for storage but also to make stored items easy to get to can make working in the kitchen a pleasure and will add more than their cost to the value of the home.

Removal of the old cabinets ordinarily requires only the removal of the woodscrews holding them to the wall or bulkhead over the cabinets. Woodscrews securing the new cabinets must be placed where they will encounter studs or solid framing. The prices quoted for new cabinets are usually installed prices; you may find that the saving by doing the job yourself is not really worthwhile.

With prices of new homes going out of sight, the advantages of many home improvements, as for instance the installation of new kitchen cabinets, become more apparent with every passing day.

Fig. 183   Visible storage in a tall cabinet provides an efficient adjustable system.

Fig. 184   Multistorage pantries are designed with adjustable shelves.

Courtesy Quaker Maid, a Tappan Division

Fig. 185   This new kitchen features a pull-out tabletop.

Fig. 186   This cabinet contains a double swing-out, bringing pans and groceries to fingertips.

Fig. 187   This glide-out wastebasket eliminates a wastebasket in the corner.

Chapter Forty-Four

The Whole-House Ventilator

Courtesy NuTone Division

Fig. 188   Cool air is pulled in through the doors and windows by the whole-house ventilator, forcing hot air out through attic louvers.

247

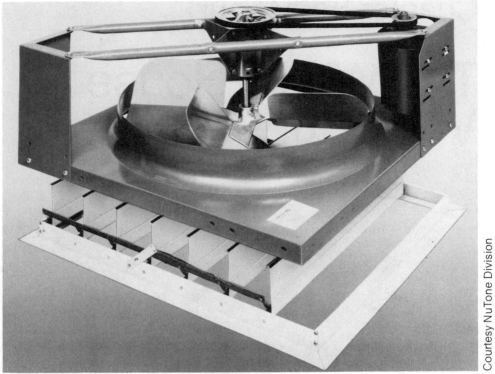

Fig. 189   The whole-house ventilator is mounted above louvers, in the
            attic.

A whole-house ventilator can greatly reduce the cost of air conditioning (cooling). With central air conditioning three electric motors
are in operation: the compressor motor, the air-handling fan motor
and the condenser fan motor; at times all three are in operation at
the same time.

　　　A whole-house ventilator is an attic fan mounted above a
louver (Fig. 189), located in the ceiling of the top floor. The house
shown in Fig. 188 is a two-story home; the fan will function equally
well in a single-story home.

　　　For those who do not like air conditioning, as well as those
who do but do not like the expense of air conditioning, the whole-
house ventilator is the perfect answer. The whole-house ventilator
provides energy-saving cooling used in conjunction with air conditioning or without. Installed in a central hallway ceiling, the quiet
ventilator pushes hot air out of the house while pulling fresh cool
air in through open windows. The ventilator can be used instead of
air conditioning in late spring and early and late summer.

　　　The louvers are made with either white enamel or aluminum
finish and the heavy-duty, rugged fan is permanently lubricated.

Chapter Forty-Five

A Residential Intruder—Fire Alarm System

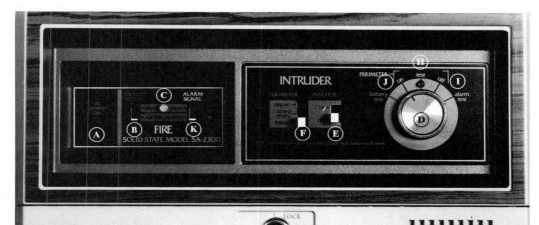

NuTone's simple controls make security easy to live with:

A. Green light lets you know AC power is on.

B. Trouble Signal Buzzer alerts you if a malfunction occurs in the Fire Detection Circuit. Can be switched off until problem is corrected.

C. Trouble Light flashes if you silence Fire Circuit Trouble Buzzer, stays on until problem is corrected and you switch it off.

D. Rotary Control Knob allows you to turn Perimeter Intrusion Circuit on and off, and to test it, test alarms, and test optional Battery standby feature.

E. Interior Circuit Switch gives you separate control of Intruder Detectors inside your home.

F. Optional Exit/Entry Control allows you to leave and re-enter your home without setting off alarms. Normally set at "Instant".

G. Key Lock Reset. If alarm is triggered, it is latched on and cannot be turned off by an intruder. Only the homeowner can shut off alarms by operating this Key Reset or an optional hidden Remote Reset. Lock also helps protect unit against casual tampering.

H. Perimeter Circuit Test Light shows Circuit is complete and operational.

I. Alarm Test Position lets you test Fire Alarm Circuit and Alarms.

J. Optional Standby Battery unit should be tested weekly at the Control Unit.

K. Alarm Signal Switch is normally set on "audible". Used with the Fire Circuit only, it can be switched to "silent" to shut off an unwanted alarm. If it is, the Trouble Buzzer sounds in the Control Unit. To help assure it is not unintentionally switched off, it must be operated with pointed instrument.

Correcting problem in the Fire Circuit has no effect on the Intruder Circuit, assuring you of continued protection.

L. Self-contained Horn Alarm.

Fig. 190   Residential intruder–fire alarm system.

The intruder–fire alarm system shown in Fig. 190 is designed for residential use. It is a built-in system with solid-state circuitry and is UL-listed. The master control has a self-contained alarm horn and the system is designed to operate on battery power in the event of power failure. A fire-detection circuit monitors heat sensors and smoke detectors and a perimeter intruder circuit guards doors and windows. Any tampering with detectors or break-in wiring causes alarms to sound. An interior intruder circuit allows the family to move freely throughout the home during the day since this circuit is normally armed only at night.

The security afforded by this type of system can make installation very desirable. Complete, easy-to-understand installation instructions are furnished with the unit.

4
GENERAL INFORMATION

General Information

THE IMPORTANCE OF KEEPING RECORDS

Nearly everything we buy is covered by some type of warranty or guarantee. These guarantees will be needed if an item proves defective or requires servicing. Start right off by keeping all warranties and guarantees (and other pertinent information) in alphabetically indexed folders in a file. If you buy a new home, the builder should supply you with a list of all subcontractors who supplied and installed material for the home. If at some future time you need a part for the cabinet work or some matching floor tile, you will know whom to contact. Get in the habit of filing and keeping records of purchases of household equipment; these guarantees often will cover the cost of replacement at no cost, except for a small labor charge, of equipment such as water heaters and garbage disposers if this equipment fails during the warranty period. In addition, on a purchase of a new house, ask the builder to leave you a bundle of the type and color of the roofing shingles used on your home. Keep these shingles in a cool place for replacements if you lose some shingles in a storm. You should also ask the builder to leave a few pieces of wood parquet flooring, squares of vinyl tile, paint used in construction or any other material that may be hard to match in years to come.

KITCHEN VENT HOOD FANS

The fan in the vent hood over the kitchen stove serves a very useful purpose by removing steam, smoke and odors from the kitchen. A removable screen designed to trap grease is mounted in the range hood below the fan. This screen should be removed once a month and washed in hot soapy water to remove the grease, rinsed well and shaken dry. A certain amount of grease will get by the screen and onto the fan blades and into the fan motor. The exhaust duct from a vent hood should extend above the roof line, and in cold weather cold air will drop down the duct and harden the grease present in the fan motor. Duct fans usually have four speed settings: if the fan is turned to a low speed setting to start in cold weather the hardened grease may prevent the fan from starting and cause the motor windings to burn up. During

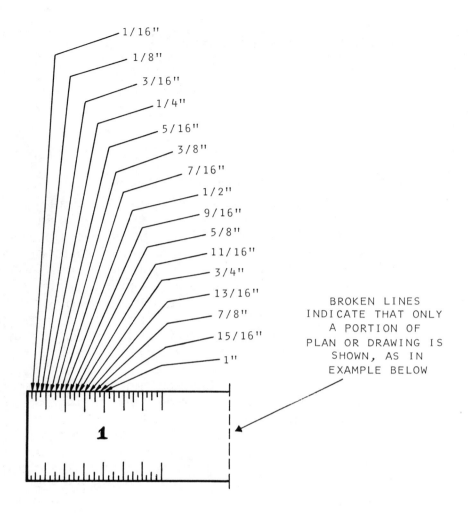

1/16"
1/8"
3/16"
1/4"
5/16"
3/8"
7/16"
1/2"
9/16"
5/8"
11/16"
3/4"
13/16"
7/8"
15/16"
1"

BROKEN LINES
INDICATE THAT ONLY
A PORTION OF
PLAN OR DRAWING IS
SHOWN, AS IN
EXAMPLE BELOW

X- AS IN 2" X 4" MEANS BY (2" BY 4")

THE SYMBOL " MEANS INCHES--2"

THE SYMBOL ' MEANS FEET--2'
2'-2" = 2 FT. AND 2 IN.

d IS THE SYMBOL FOR PENNY, USED IN NAIL SIZES

KW = KILOWATTS- 1000 WATTS

KW/hr= KILOWATTS PER HR.   = ELECTRICAL
                                     GROUND
Cf/m = CUBIC FEET PER MINUTE

Gpm = GALLONS PER MINUTE

MIP = MALE IRON PIPE   FIP = FEMALE IRON PIPE

Fig. 191   Helpful information.

cold weather start a hood vent fan on *high speed*; allow it to run a few minutes to warm the motor; then the fan can be switched to a lower setting if desired.

The fan can be operated continuously at low speed during cold weather if desired.

CLEAN THE FIREPLACE CHIMNEY

Fireplace chimneys should be cleaned every year before the start of the heating season. Every type of wood will leave some residue of creosote and tars in the flue liner of the chimney. The heavier the buildup of creosote and tars, the more likelihood there is that a fire could start in the chimney. Chimney fires can get *extremely* hot, and are dangerous because this type of fire can cause the flue liner to crack, exposing the bricks surrounding the liner to the intense heat, causing them to crack and let the fire out into the attic space. Don't let your chimney become a fire hazard. You can clean the chimney, assuming you can get on the roof and to the chimney top, by putting one or two bricks in a burlap sack, tying the sack securely to a good strong rope and then getting on the roof and dropping the sack down into the chimney and working it up and down to dislodge tars, soot, etc. The fireplace opening should be blocked with cardboard or old carpeting to keep the dislodged soot from getting out into the room. And of course never try to clean the chimney with a fire going in the fireplace.

Fireplace shops and hardware stores sell special chemicals which can be thrown on the fire while it is burning to help keep the chimney clean. There are also firms which specialize in cleaning chimneys and advertise their services in the yellow pages of the phone book.

The most common cause of chimney fires is too large a fire in the fireplace or burning paper or cardboard in the fireplace. If your chimney should catch fire, *first call your fire department*, then thoroughly soak an old blanket or piece of carpet with water and cover the front of the fireplace. The wet blanket will keep air from getting into the fireplace and the chimney fire should starve from lack of oxygen.

OVERHEAD-DOOR MAINTENANCE

The rollers and tracks of overhead doors should be greased once a year. Wheel-bearing grease should be dabbed on the inside of the tracks. A small amount at one place on the track will do the job: the rollers will pick up the grease and spread it over the entire track. Certain types of doors have tightly coiled springs mounted on a horizontal shaft directly over the opening. Springs of this type should be oiled at the same time the tracks are greased. Use a pump-type oil can and apply a small stream of number 30 oil to the top of the entire length of the springs.

FIRE EXTINGUISHERS

Every home should have at least two fire extinguishers, located where they are easily accessible—one in the kitchen and one in the garage. If your home has a basement an extinguisher should be within easy reach of the stairway. Quick use of an extinguisher on a small fire can prevent it from becoming a large one. If a fire has a good start, call the fire department first; don't try to extinguish the fire until you've called for help.

Dry chemical extinguishers are effective on Class A, B and C fires and are the recommended type for home use.

> *Class A fires:* Burning paper, wood or fabric where water or multipurpose dry chemical is the most effective agent.
> *Class B fires:* Burning flammable liquids. Smothering is required.
> *Class C fires:* In electrical equipment. The extinguishing agent must be a nonconductor of electricity.

Fire extinguishers should be checked periodically and should be inspected and recharged once a year. Inspection and recharging services are usually provided by the fire department.

SMOKE DETECTORS

Many building codes now require that smoke detectors be installed in new homes, and many home insurance companies quote lower rates if smoke detectors are installed. If you do not already have smoke detectors installed in your home, their installation is strongly recommended. Nearly all types sense not only smoke but also the presence of abnormal heat and certain gases. Some models also have a light which can help to guide occupants to safety if the alarm is triggered.

All types of smoke detectors should be tested periodically to ensure that they are in working order.

Smoke detectors should be placed in hallways outside of bedrooms and in an area near the kitchen. If a smoke detector is placed *in* a kitchen, smoke or steam connected with normal cooking may trigger it. Smoke detectors are also recommended in the garage and in the laundry room.

Smoke detectors are packaged with mounting hardware and can be quickly and easily installed.

WALL FASTENERS

Wall fasteners or anchors are used to fasten or secure objects such as picture hangers, lighting fixtures or brackets to a wall, ceiling or floor. The three most widely used kinds are Molly bolts, toggle bolts and plastic shields.

Molly bolts can be used in any surface having a hollow interior close to the outside surface. A Molly is installed by drilling a hole slightly larger than the outside body of the Molly, inserting it into the hole and tightening the bolt. Teeth on the flange or outside face of the Molly grip the wall surface and prevent the Molly from turning when the bolt is tightened. As the bolt is tightened the Molly body expands, securing it to the wall. The bolt can then be removed, inserted through the object to be secured and tightened back into the molly. Molly bolts are made in several sizes and lengths, depending on the application. The larger the Molly the more strength it possesses. Mollys are suitable for use in gypsum drywall, plaster walls or hollow concrete block.

Toggle bolts are also used to secure objects to a surface (walls, floors, ceilings) with a hollow interior close to the outside surface. They are very strong when properly installed. A toggle bolt requires a hole large enough to insert the folded toggle as shown in Fig. 192. When the toggle enters the hollow it opens and as the bolt is tightened the toggle grips the inside surface. There is a marked difference in the way a toggle bolt and a Molly bolt are used. A Molly bolt can be removed, leaving the body of the fastener in place. When a toggle bolt is inserted into the hole, the toggle opens and if the bolt is then removed the toggle will fall off and cannot be used again. For this reason when a toggle bolt is used, the bolt must be inserted through the object to be secured and then the bolt can be inserted into the hole and tightened.

Plastic shields are made for use in solid surfaces. Kits containing the shields, metal screws and the correct size of masonry drill bit are sold in hardware and building-supply stores. An electric drill motor and the drill bit are used to drill a hole slightly deeper than the length of the plastic shield. Blow any dust or loose material from the hole, then insert the plastic shield,

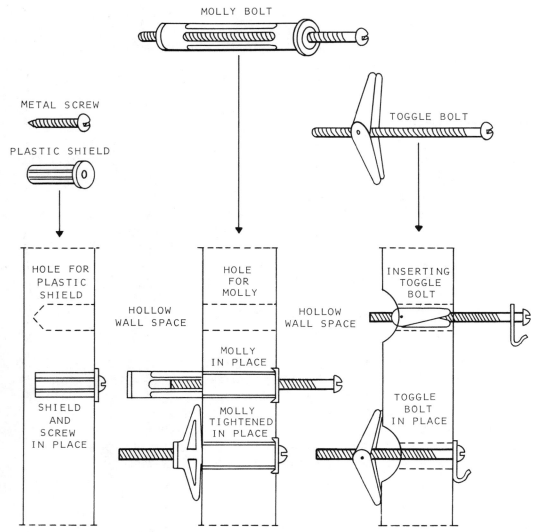

MOLLY BOLT

METAL SCREW

PLASTIC SHIELD

TOGGLE BOLT

HOLE FOR
PLASTIC
SHIELD

HOLLOW
WALL SPACE

SHIELD
AND
SCREW
IN PLACE

HOLE
FOR
MOLLY

HOLLOW
WALL SPACE

MOLLY
IN PLACE

MOLLY
TIGHTENED
IN PLACE

INSERTING
TOGGLE
BOLT

TOGGLE
BOLT
IN PLACE

Fig. 192   Three types of wall fastener.

tapping it into place with the hammer if necessary. Insert the metal screw through the object to be secured and tighten the screw. When properly installed this type anchor will hold an object securely.

NAILS

The nails shown in Fig. 193 are drawn to actual size to help you determine which nails are needed for a project. These are common nails, some other kinds are: box nails, finishing nails, cut nails and concrete nails. Common nails are available in either plain steel, galvanized steel or aluminum. When heads of nails will be

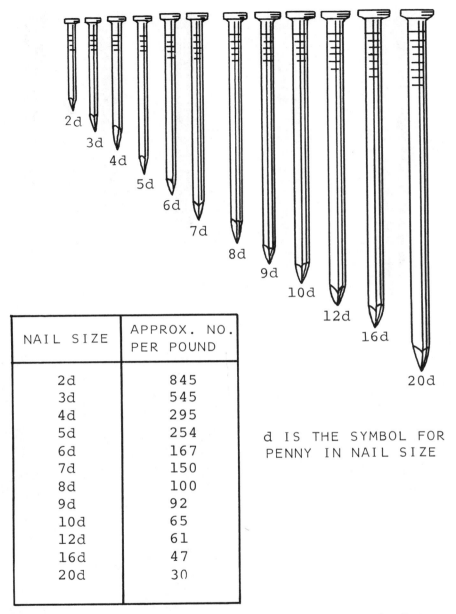

NAIL SIZE	APPROX. NO. PER POUND
2d	845
3d	545
4d	295
5d	254
6d	167
7d	150
8d	100
9d	92
10d	65
12d	61
16d	47
20d	30

d IS THE SYMBOL FOR
PENNY IN NAIL SIZE

Fig. 193   Actual sizes of common nails and numbers of nails per pound.

exposed, as when applying wood lap siding, galvanized or aluminum nails should be used to prevent rust streaks due to
weathering.

Box nails are slightly smaller in diameter than common
nails and are less likely to cause the wood being nailed to split.

Finishing nails have a small head which can be driven
below the surface by using a nail set and the nail will not show.

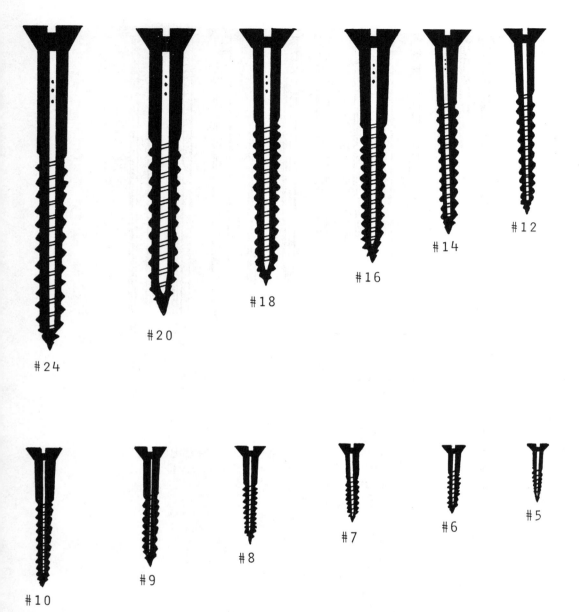

#24

#20

#18

#16

#14

#12

#10

#9

#8

#7

#6

#5

Fig. 194   Sizes and numbers of woodscrews.

The small hole made by the nail head can be filled with wood filler, putty or plastic wood and then finished with any desired finish.

Cut nails or concrete nails are hardened and are used primarily to nail wood to concrete surfaces.

When you start a project requiring nails, your building material supplier can advise you as to the proper kind and size nails to use.

Some kinds of wood will tend to split when being nailed. When you run into this problem, place the head of the nail on a hard solid surface and use your hammer to blunt the nail point. A blunt-headed nail is much less apt to split the wood.

WOODSCREWS
The flathead woodscrews shown full size in Fig. 194 should help you determine the size needed. Woodscrews are also made with round and oval heads.

When woodscrews are used in "hard" woods such as oak, cherry, walnut, beech and maple a pilot hole one-half the diameter of the screw being used should be drilled to permit easy driving of the screw. The pilot hole will also prevent the wood from splitting. Rubbing paraffin or beeswax on the screw threads will also aid in inserting the woodscrew.

Don't throw that old file away just because the teeth are full of filings. Flatten a large nail or a thin piece of steel and clean the teeth, as shown in Fig. 195. The file teeth will cut the grooves in the cleaning tool as the teeth are being cleaned.

Fig. 195   How to make a file-cleaning tool.

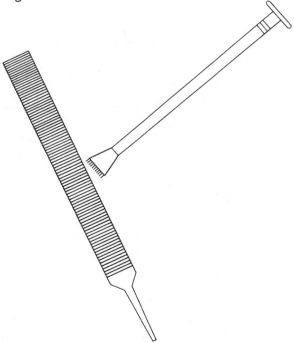

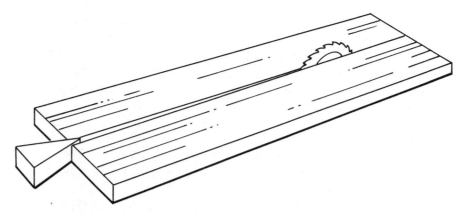

Fig. 196   A wedge will keep saw blade from binding.

When a table saw or a radial arm saw is ripping a long board, the cut often closes behind the blade, causing the blade to bind. A wedge driven into the saw cut, as shown in Fig. 196, will keep the cut open and prevent binding.

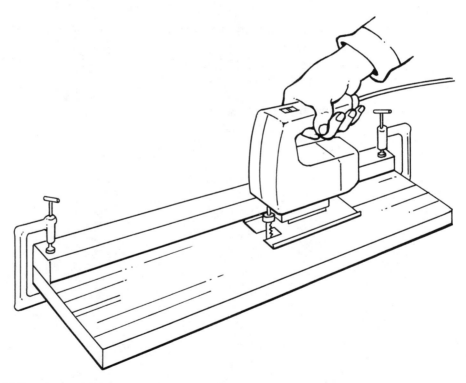

Fig. 197   A straightedge used as a guide.

A straightedge clamped to the board being cut will serve as a guide for the shoe of a sabre saw, and will ensure a straight cut if the saw is held against it, as shown in Fig. 197.

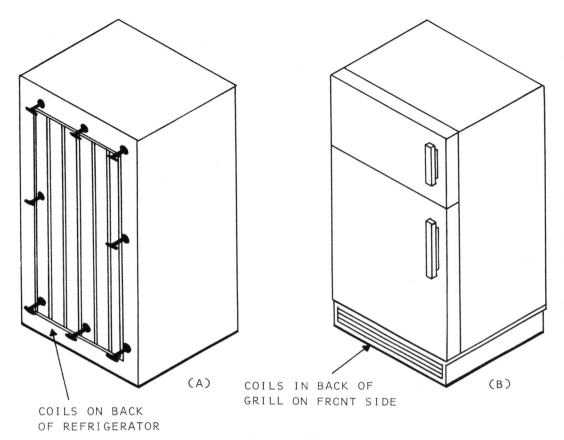

(A)

COILS ON BACK
OF REFRIGERATOR

COILS IN BACK OF
GRILL ON FRONT SIDE

(B)

Fig. 198   Cleaning condenser coils on refrigerators.

Heat is extracted from the air inside a refrigerator and dissipated through coils either on the back of a refrigerator (A) or below it, behind a grill on the front (B), Fig. 198. For both economical and efficient operation these coils should be kept free from dirt, dust and lint. They should be cleaned as recommended by the manufacturer at least once a month.

Index

Note: Page references in italics indicate figures not immediately adjacent to related text.